AF684575

SCIENCE by DESIGN

CONSTRUCT a BOAT, CATAPULT, GLOVE, and GREENHOUSE

SCIENCE by DESIGN

CONSTRUCT a BOAT, CATAPULT, GLOVE, and GREENHOUSE

Developed by TERC

TERC

NSTA press
National Science Teachers Association
Arlington, Virginia

NSTApress®
National Science Teachers Association

Claire Reinburg, Director
Jennifer Horak, Managing Editor
Andrew Cooke, Senior Editor
Wendy Rubin, Associate Editor
Agnes Bannigan, Associate Editor
Amy America, Book Acquisitions Coordinator

Art and Design
Will Thomas Jr., Director
Rashad Muhammad, Graphic Designer, Cover and Interior Design

Printing and Production
Catherine Lorrain, Director

National Science Teachers Association
Gerald F. Wheeler, Executive Director
David Beacom, Publisher

1840 Wilson Blvd., Arlington, VA 22201
www.nsta.org/store
For customer service inquiries, please call 800-277-5300.

Copyright © 2013 by the National Science Teachers Association.
All rights reserved. Printed in the United States of America.
16 15 14 13 4 3 2 1

This curriculum was developed by TERC, Cambridge, Massachusetts. Funded in part by a grant from the National Science Foundation. This book was prepared with the support of the National Science Foundation under Grant Nos. ESI-9252894 and ESI-9550540. However, any opinions, findings, conclusions, and/or recommendations herein are those of the authors and do not necessarily reflect the views of the National Science Foundation.

NSTA is committed to publishing material that promotes the best in inquiry-based science education. However, conditions of actual use may vary, and the safety procedures and practices described in this book are intended to serve only as a guide. Additional precautionary measures may be required. NSTA and the authors do not warrant or represent that the procedures and practices in this book meet any safety code or standard of federal, state, or local regulations. NSTA and the authors disclaim any liability for personal injury or damage to property arising out of or relating to the use of this book, including any of the recommendations, instructions, or materials contained therein.

Permissions
Book purchasers may photocopy, print, or e-mail up to five copies of an NSTA book chapter for personal use only; this does not include display or promotional use. Elementary, middle, and high school teachers may reproduce forms, sample documents, and single NSTA book chapters needed for classroom or noncommercial, professional-development use only. E-book buyers may download files to multiple personal devices but are prohibited from posting the files to third-party servers or websites, or from passing files to non-buyers. For additional permission to photocopy or use material electronically from this NSTA Press book, please contact the Copyright Clearance Center (CCC) (*www.copyright.com*; 978-750-8400). Please access *www.nsta.org/permissions* for further information about NSTA's rights and permissions policies.

SCILINKS. THE WORLD'S A CLICK AWAY *Featuring SciLinks—a new way of connecting text and the Internet. Up-to-the minute online content, classroom ideas, and other materials are just a click away. For more information go to* www.scilinks.org/Faq.aspx.

Library of Congress Cataloging-in-Publication Data
Science by design : construct a ... boat, catapult, glove, and greenhouse / developed by TERC.
pages cm
ISBN 978-1-936959-48-8 (print) -- ISBN 978-1-936959-57-0 (e-book) (print) 1. Science--Study and teaching (Secondary)--Activity programs. 2. Technology--Study and teaching (Secondary)--Activity programs. I. Technical Education Research Centers (U.S.)
Q181.S3597 2013
507.1'2--dc23
2012033845

CONTENTS

Part II: Construct-a-CATAPULT

Part III: Construct-a-GLOVE

Part IV: Construct-a-GREENHOUSE

Preface

The original Science by Design series was published in four volumes in 2000 by NSTA Press. The development of the series was funded by the National Science Foundation, and each of the four units that constituted the original volumes in the series—Construct-a-Boat, Construct-a-Catapult, Construct-a-Glove, and Construct-a-Greenhouse—were created by TERC, a nonprofit education research and development organization.

The original four volumes have been updated and combined into a single book, *Science by Design*.

Organization of the Units

Each unit includes an introductory section with notes from the developers on integrating science and technology, schedule and cost, key ideas, inquiry and design, student portfolios, standards and benchmarks connections, SciLinks, and a course outline. The Introduction is followed by five activities:

1. Design Brief
2. Quick-Build
3. Research
4. Development
5. Communication

Compare these materials to a highway: if you rush straight through, your students will learn only a little about the territory they have crossed. We provide a number of interesting side roads (see Appendix A of each unit), which you may or may not choose to investigate. Following these side roads will take more time, but there is no better way to explore the linkage of inquiry and design. For your first trip, you may want to stay close to the highway, but as you gain experience, we hope you will travel farther and farther from it.

Inquiry and Design

Students undertake inquiry and design as iterative, multidisciplinary processes through which students develop abilities in designing and conducting investigations, recognizing and applying models, constructing explanations, making predictions, creating solutions, decision making, building, testing, and evaluating.

The Inquiry Process

The inquiry process is often viewed as a cycle of action that repeats until the investigators reach a satisfying solution. It can be described with seven basic elements: identify, plan, research, experiment, explain, evaluate, and communicate. Appendix A of each unit includes a detailed description of the inquiry process and a list of questions related to the process.

The Design Process

The design process is often viewed as a cycle of action that repeats until the designers reach a satisfying solution. It can be described with seven basic elements: identify, create, investigate, choose, implement, evaluate, and communicate. Appendix A of each unit includes a detailed description of the design process and a list of questions related to the process.

PREFACE

Standards and Benchmarks

The Introduction of each unit contains a table linking the tasks of that unit to the following standards and benchmarks:

- *Project 2061: Benchmarks for Science Literacy* (American Association for the Advancement of Science [AAAS] 1993)
- *Technology for All Americans: A Rationale and Structure for the Study of Technology* (International Technology Education Association [ITEA] 1996)
- *National Standards for Social Studies Teachers* (National Council for the Social Studies [NCSS], Task Force on Social Studies Teacher Education Standards 1997)
- *Professional Standards for Teaching Mathematics* (National Council for Teachers of Mathematics [NCTM] 1991)
- *National Science Education Standards* (NSES) (National Research Council 1996)

The complete reference for each of these sources is given at the end of the table in each unit.

Text Reconstruction

Appendix B of each unit provides text reconstruction exercises that cover the key ideas of the unit. This well-established technique for reading and writing improvement has roots going back to Benjamin Franklin as well as a number of famous authors. Many teachers find that including text reconstruction in a reading assignment highly motivates students and results in a much higher rate of homework completion. Chapter 5 of *Why Johnny Can't Write* (Linden 1990) includes complete instructions on how to design your own exercises. Additional exercises can be found in *Analyze, Organize, Write* (Whimbey and Jenkins 1987).

An instructional process that uses inquiry or design places great demands on class time. It is impossible to cover all essential content within the few hours per week students spend in class. Science and technology courses must therefore insist that students learn from reading. This means they must also provide realistic opportunities for students to improve their ability to learn from reading.

Text reconstruction works as a method of improving reading skills by focusing student attention on the most important elements of the reading task. First, it changes the reader's perception of his or her role from that of a passive absorber of information to that of an active agent who must sort out a puzzle. This is probably the main reason text reconstruction exercises are popular with students. Second, text reconstruction forces students to pay attention to the logic of a paragraph. In science, it is not the separate ideas that are important, but rather the logic that ties them together. When passively reading a paragraph, one can easily miss that logic, but in text reconstruction it is impossible to complete the task without thoroughly understanding these interconnections. A student who reconstructs a paragraph will understand its structure and meaning far more deeply than a student who memorizes every word but considers them only in their current order.

In *Science by Design*, we employ various techniques to encourage student reading and writing. These are not extras, but rather essential elements of the program. The exercises assume that your students are relatively strong readers. If your students are weak readers or are not used to serious homework, then you will need to increase the amount of attention you pay to improving reading skills. Text reconstruction can be used to convert any kind of reading assignment into a stimulating puzzle. The more use you make of text reconstruction, the more your students will read and the better they will understand.

Assessment and Portfolios

Student activity sheets may be used for formative or summative assessment. The first Snapshot of Understanding in each unit is intended as a prelearning index of prior knowledge. The answers in this Snapshot may be compared with the answers in the final Snapshot given at the end of the last activity, for student self-assessment of learning. Because group work is stressed throughout each unit, group assessment may prove to be more appropriate than individual scores. However, depending on your class objectives, homework assignments may provide the best measure of individual performance.

A portfolio can be a useful tool for maintaining individual accountability in a teamwork environment because it allows students to capture representative samples of their work over time. One resource among the many guides to portfolio assessment is *Portfolio Assessment: A Handbook for Educators* by James Barton and Angelo Collins (1997).

Internet Searches

Internet searches are often an important part of inquiry units. When assigning an internet search, a few cautionary words about censorship, commercialism, and privacy issues should be included. You may want to consult your school's computer lab director for assistance.

The following suggestions are intended to improve the educational effectiveness of internet searches:

- Set an objective, assignment, or contest that requires each student or pair of students to turn in a search path or research result printout at the end of the session. Inform students how the deliverable will be used for assessment.
- Schedule enough time for the activity so that the slower navigators will be able to reach predetermined destinations, while offering options for speedier students to move beyond those destinations).
- Consider pairing students—inexperienced with familiar, confident with scared, leaders with followers, disciplinary problems with angels, and so on—to create peer teaching opportunities and reduce individual distractions. Pairing will also decrease the network traffic in your lab, which could otherwise decrease computer response times.
- It is useful to review the structure of search commands and how to limit a search by combining key words. You might want to prepare a list of key words that you have tried and found productive for groups to pick from according to their interests or assigned objective. Having everyone achieve a common destination and then diverge will increase the class's

overall depth of exploration if results are to be shared.

- After the search, you may want to provide a list of websites from your own resource list and invite students to submit additions to it. In addition, you may want to ask students to bookmark useful websites.

References

Barton, J., and A. Collins. 1997. *Portfolio assessment: A handbook for educators.* Addison-Wesley.

Linden, M. J. 1990. *Why Johnny can't write: How to improve writing skills.* Hillsdale, NJ: Erlbaum.

Whimbey, A., and E. L. Jenkins. 1987. *Analyze, organize, write:* A structured program for expository writing. Revised ed. Hillsdale, NJ: Erlbaum.

Suggested Readings

Bennett, K. E., and W. C. Ward. 1993. *Construction versus choice in cognitive measurement: Issues in constructed response, performance testing, and portfolio assessment*. Hillsdale, NJ: Erlbaum.

Germann, P. J., S. Haskins, and S. Auls. 1996. Analysis of nine high school biology laboratory manuals: Promoting scientific inquiry. *Journal of Research in Science Teaching* 33 (5): 475–499.

Gitomer, D. H. 1988. Individual differences in technical troubleshooting. *Human Performance* 1 (2): 111–131.

Gokhale, A. 1997. Writing in the technology discipline. *The Technology Teacher* 56 (8): 11–23.

Haas, N., and C. Boston. 1994. One school experiments with performance-based assessments. *The Eric Review* 3: 13–14.

Keys, C. W. 1995. An interpretive study of students' use of scientific reasoning during a collaborative report writing intervention in ninth grade general science. *Science Education* 79 (4): 415–435.

Naveh-Benjamine, M., and Y. Lin. 1991. *Assessing student's origination of concepts: A manual for measuring course-specific knowledge structures*. Ann Arbor: University of Michigan, National Center for Research to Improve Post-secondary Teaching and Learning.

Raizen, S. A., P. Sellwood, R. D. Todd, and M. Vickers; National Center for Improving Science Education. 1995. *Technology education in the classroom: Understanding the designed world*. San Francisco: Jossey-Bass.

Roth, W. M., and G. M. Bowen. 1995. Knowing and interacting: A study of culture, practices and resources in a grade 8 open-inquiry science classroom guided by a cognitive apprenticeship metaphor. *Cognition and Instruction* 13 (1): 73–128.

Rudner, L., and C. Boston. 1994. Performance assessment. *The Eric Review* 3: 2–12.

Wong, E.D. 1993. Understanding the generative capacity of analogies as a tool for explanation. *Journal of Research in Science Teaching* 30 (10): 1259–1272.

Acknowledgments

The creation of the NSTA Science by Design series builds on five years of research and development at TERC. This work was funded by two National Science Foundation grants, ESI-9252894 and ESI-9550540, and was directed by John Foster, David Crismond, William Barowy, and Jack Lochhead.

We are especially grateful for the vision, guidance, and prodding of our program officer, Gerhard Salinger, who was our GPS in uncharted territory.

We were helped in innumerable ways by an especially insightful advisory board: Joan Baron, Goery Delacote, Andy DiSessa, Woodie Flowers, John Foster, Mike Hacker, Colleen Hill, Gretchen Kalonji, Robert McCormick, Jim Minstrell, Jim Neujahr, and David Perkins.

The members of the TERC team who contributed to the development and testing of the Science by Design units included Tim Barclay, William Barowy, Cathy Call, Judith Collison, David Crismond, Brian Drayton, Christine DiPietrantonio, Joni Falk, John Foster, June Foster, Riley Hart, Nathan Kimball, Felicia Lee, Jack Lochhead, Tasha Morris, Tracy Noble, Alison Paddock, Meghan Pfister, Lee Pulis, Jerry Touger, Margaret Vickers, Paul Wagoner, Kelly Wedding, and Amy Weinberg.

The following hard-working consultants added greatly to our efforts: Hilton Abbott, Robert B. Angus, Carol Ascher, Warren R. Atkinson, Earl Carlyon, Michael Clarage, Jan Hawkins, Kathy Kittredge, Crispin Miller, James E. LaPorte, Kjell-Jan Rye, Rick Satchwell, Mike Stevens, and Ron Todd.

Special mention must be made of the enthusiasm, dedication, and long hours contributed by David Crismond and Earl Carlyon.

The authors of the final Science by Design units were William Barowy (Boat), Felicia Lee (Greenhouse), Jack Lochhead, Alison Paddock, and Lee Pulis (Catapult and Glove) of TERC.

The original four volumes of the Science by Design series were produced by NSTA Press: Shirley Watt Ireton, director; Beth Daniels, managing editor; Erin Miller, associate editor; Jessica Green, assistant editor; Michelle Treistman, assistant editor; Anne Early, editorial assistant. Beth Daniels was project editor for the Science by Design series.

Field testing of the Science by Design series depended on the dedication of dozens of teachers and the helpful cooperation of their school systems: Jerian Abel, Northwest Regional Education Laboratory; Bruce Andersen, Buker Middle School; Dave Armstrong, Lawrence Middle School; Henry Bachand, Mansfield High School; Hilda Bachrach, Dana Hall School; Ronald Bjorklund, Leicester High School; Marcella Boyd, Manchester Junior High School; Karen Bouffard, Governor Dummer Academy; Althea Brown, Medford High School; Lee Burgess, Lawrence Middle School; David Corbett, Whittier Regional Vocational Technical High School; Steve Cremer, Braintree High School; Deborah Crough, Long Beach High School; Ron Daddario, McCall Middle School; Raymond P. Gaynor, Reid Middle School; Elizabeth George, Westborough High School; Pam Glass, Talbot Middle School; Rick Harwood, Ware High School; Gary Herl, Tantasqua Regional Junior High School; Kate Hibbitt, Lincoln School; Patrick Keleher, Norwood Junior High School; Marty Kibby, Minetron Technology Education Office; Matias Kvaternik, Chelmsford Public

ACKNOWLEDGMENTS

Charter School; Jeff Leonard, F.A. Day Middle School; Walter Lewandowsler, Bartlett High School; David R. Littlewood, Agawam Junior High School; John Matthews, Southwick Tolland Regional High School; Eilen McCormack, South Junior High Brockton; Scott McDonald, Needham High School; Brian McGee, Lexington Middle School; Bob Meltz, Manchester Junior-Senior High School; Charles O'Reilly, Lexington High School; Constance Patten, Lincoln-Sudbury High School; Fred Perrone, East Junior High School; Joe Pignatiello, Somerville High School; Doug Prime, Lancaster Middle School; Michael Rinaldi, Bedford High School; Thomas Rosa, Walsh Middle School; Eugene A. Santoro, Silver Lake Regional High School; John Schott, Smith Academy; Bruce Seiger, Wellesley High School; Douglas Somerville, Woodword Middle School; John Stamp, Manchester High School; Mike Stevens, Maynard High School; Michael Sylvia, Charles E. Brown Middle School; Syd Taylor, Mahar Regional School; Ted Vining, Monument Regional High School; and Frank Viscardi, Framingham High School.

This project would not have been possible without the help and critique of hundreds of students, whom we regretfully cannot mention by name.

Part I

CONSTRUCT a BOAT

Part I

Introduction

Integrating Science and Technology

Construct-a-Boat is aligned with the National Science Education Standards for process and content standards in both physical science and mathematics, as shown in the Standards and Benchmarks Connections table later in this section. Through inquiry and design, students develop conceptual understanding of electromechanical energy transfer, friction, and mathematical modeling. Because design activities motivate inquiry, and inquiry informs design, students engage in the iterative processes of scientific inquiry and technological design through a variety of hands-on activities.

Schedule and Cost

The minimum time needed to complete the core unit is about 12 class sessions. More time will be needed if you choose to extend the unit either by undertaking a more advanced design and fabrication process or by pursuing more advanced treatment of mathematical modeling.

Working in teams, students make a quick-build model boat using the Design Brief challenge and instructions (1 or 2 class sessions). During the research and development phases (at least 7 class sessions), students take baseline measurements, identify relevant variables, design and conduct experiments, study the model boat system, propose conceptual models, and generate possible solutions to their design problems. Students develop designs, build the models, conduct further investigations, analyze their data, and redesign their models if necessary. The unit concludes with student presentations of products, including supporting scientific arguments (3–4 class periods).

We suggest that you impose a very low cost limit on materials ($1–5 per student) and encourage the use of recycled materials.

Key Ideas

Each key science idea used in Construct-a-Boat is covered in a text reconstruction exercise in Appendix B (p. 82).

Forces, Speed, and Acceleration

As they seek to meet the design specifications of the boat, students learn about mass, speed, acceleration, and forces through practical application of these concepts. Students investigate the relationships among these variables through qualitative observation and quantitative measurement of changes and rates of change in each variable.

Systems

Students work with variables to study system behavior. They learn to construct the feedback loops that determine the limits of model boat performance.

Modeling

Students apply the science and math necessary to build an accurate scale model, and they extend this analysis to computer modeling. Homework and class assignments guide students through problem solving in algebra and in plane and solid geometry, and they show students how to make connections to other disciplines.

Inquiry and Design

Because the focus of Construct-a-Boat is on the design-and-build process, students develop conceptual, mathematical, and computer models with structured guidance, and they apply expert-built models to the making of a scale model. Optional homework readings on the history of boat-building provide context for the technological design background given to the students in this unit.

Student Portfolios

The following items can be accumulated in portfolios for summative assessment:

- Pretest: Snapshot of Understanding
- Initial questions: Design Brief
- Individual information search
- Sketch of boat hull design
- Brainstorming record
- List of variables
- Research and results
- Group process description: Inquiry Process
- Group process description: Design Process
- Prototype demonstration notes
- Group summary documentation: Product Prospectus
- Posttest and self-assessment: Snapshot of Understanding

Part I

INTRODUCTION

Standards and Benchmarks Connections

Task	Source
Students recognize the evolution of computer models and scale models in the shipbuilding industry and the impact of resulting performance improvement on the environment. **Standard/Benchmark:** Science, Technology and Society	NCSS VIII
Students recognize that systems have layers of controls. **Standard/Benchmark:** Design and Systems	AAAS 9–12
Students troubleshoot common mechanical and electrical systems, checking for possible causes of malfunction. **Standard/Benchmark:** Manipulation and Observation	AAAS 9–12
Students recognize the use of scale modeling for performance testing. **Standard/Benchmark:** Design and Systems	AAAS 9–12
Students use tools safely for construction. **Standard/Benchmark:** Manipulation and Observation	AAAS 9–12
Students identify and describe variables that affect the efficiency of the boat as a mechanical system. **Standard/Benchmark:** Systems; Processes: Determining and Controlling Behavior of Technological Systems; Evidence, Models, and Explanation	AAAS 9–12, ITEA II, NSES K–12
Students interpret scale drawings, interpret and draw three-dimensional objects. **Standard/Benchmark:** Communication Skills; Geometry From a Synthetic Perspective	AAAS 9–12, NCTM 7
Students understand how things work and design solutions using systems analysis. **Standard/Benchmark:** Systems	AAAS 9–12
Students compare model predictions to observations and use computer models to explore the logical consequences of a set of instructions. **Standard/Benchmark:** Design and Systems; Information Processing	AAAS 9–12
Students recognize that different properties are affected to different degrees by changes in scale. **Standard/Benchmark:** Scale	AAAS 9–12
Students understand how mathematical modeling aids in technological design by simulating how a proposed system would theoretically behave; students recognize the limits of a mathematical model in how well it can represent how the world works. **Standard/Benchmark:** Symbolic Relationships; Mathematics as Problem Solving	AAAS 9–12, NCTM 1

INTRODUCTION

Task	Source
Students recognize tables, graphs, and symbols as alternative ways to represent data and relationships that can be translated from one to another, and use the computer for producing tables and graphs and for making spreadsheet calculations. **Standard/Benchmark:** Symbolic Relationships; Manipulation and Observation	AAAS 9–12
Students model real-world phenomena with a variety of functions, and they represent and analyze relationships using tables, verbal rules, equations, and graphs to translate among tabular, symbolic, and graphical representations of functions. **Standard/Benchmark:** Functions	NCTM 6
Students design and conduct a scientific investigation, and they formulate and revise scientific explanations and models using logic and evidence. Standard/Benchmark: Science as Inquiry	NSES 9–12
Students observe that objects change their motion only when a net force is applied. **Standard/Benchmark:** Physical Science	NSES 9–12
Students develop abilities of technological design, including brainstorming design ideas, choosing among alternative solutions, implementing a proposed solution, and evaluating the solution and its consequences. **Standard/Benchmark:** Science and Technology	NSES 9–12
Students communicate the design problem, process, and solution, and write clear, step-by-step instructions for conducting investigations. **Standard/Benchmark:** Communication Skills; Science and Technology	AAAS 9–12, NSES 9–12
Students self-assess their learning by comparing pre- and post-Snapshots of Understanding. **Standard/Benchmark:** Issues in Technology; Science, Technology, and Society	AAAS 9–12, NCSS VIII

Source Key:

AAAS = American Association for the Advancement of Science. 1993. *Project 2061: Benchmarks for science literacy.* New York: Oxford University Press.

ITEA = International Technology Education Association. 1996. *Technology for all Americans: A rationale and structure for the study of technology.* Reston, VA: ITEA.

NCSS = Task Force on Social Studies Teacher Education Standards. 1997. *National standards for social studies teachers.* Washington, DC: National Council for the Social Studies.

NCTM = National Council for Teachers of Mathematics. 1991. *Professional standards for teaching mathematics.* Reston, VA: NCTM.

NSES = National Research Council. 1996. *National science education standards.* Washington DC: National Academies Press.

Part I

INTRODUCTION

Course Outline

Introduction

- Overview: Model Boat Design Brief
- Snapshot of Understanding
- Scale Modeling at Work

Quick-Build

- Overview: Quick-Build Model Boat
- Quick-Build Specifications
  - » Blueprint
  - » Electrical Drawings
  - » Electrical Switch Detail Drawing
- Making a Test Tank
- Exploring the Model Boat System
- Systems Modeling Introduction Homework
- Causal Relationships: Linking Variables

Research

- Overview: Research
- Baseline Measurements
- Organizing the Data
- Identifying Variables
- Scale Model Preparation Homework
- Applications of Scaling
- Scale Model Extensions Homework
- Fluid Friction Dynamics
- Minimizing Surface Area

Development

- Overview: Development
- Designer Problem
- Builder Problem
- Prototype Construction
- Evaluating Your Design

Communication

- Overview: Communication
- Creating a Project Report
- Presentation
- Reflection and Recommendations
- Snapshot of Understanding

DAY 1

ACTIVITY 1
Model Boat Design Brief

OVERVIEW: MODEL BOAT DESIGN BRIEF

In this unit you will be researching, designing, building, and improving a model boat. For purposes of this challenge, you are an employee of Quality Boat Systems (QBS), a company that designs boat hulls for carrying people and cars. Like many other boat design companies, QBS has been able to develop cost-effective, high-performance hulls by computer modeling and testing scale models. Your customers want boats that will speed up service. QBS scientists say that the way to do this is to maximize the boat's acceleration and its top speed.

Design Challenge

As a member of a product research and development team, research how to improve acceleration and top speed by redesigning the boat hull. If successful, your new design will perform better than the quick-build model.

Scope of Work

- *Quick-Build:* Make a quick-build model boat according to plans and collect baseline performance data.
- *Research:* Identify and research key design features.
- *Development:* Redesign the model hull to improve performance; test the new design to calculate percent improvement in performance; and document progress, design improvements, and tests.
- *Communication:* Write a report and present your work.

Part I

ACTIVITY 1
MODEL BOAT DESIGN BRIEF

Specifications

Salespeople have determined that QBS can sell boats to customers with the following specifications similar to the M/V *Nantucket*:

Performance Specifications

- Maximum top speed: 30 km/hr
- Maximum acceleration: 0.1 m/s^2
- Fuel consumption: ~350 l/hr
- Power twin 3,000 horsepower turbine

Physical Specifications

- Weight: 1,170 metric tons
- Length overall: 70 m
- Beam: 18 m
- Vehicle capacity: 60 midsize cars
- Passenger capacity: 1,100 (including crew)

1

SCILINKS® THE WORLD'S A CLICK AWAY

Topic: acceleration
Go to: *www.scilinks.org*
Code: CAB01

SNAPSHOT OF UNDERSTANDING

What I Already Know About Models, Systems, and Design

The unit of study you are about to begin will challenge you to design, build, and test the performance of a model boat. To meet this challenge, you will have to investigate the physics of performance and the concepts of a model boat system. Before you begin, record a sampling of what you already know by answering the questions below. This is not a test; rather, it is a series of questions about your current knowledge of key ideas in this unit. At the end of the unit, you will answer similar questions and compare what you have learned.

1. What are some of the factors that would have significant effects on the speed and acceleration of a model boat?

2. Do these factors have any effect on each other? If so, what might those effects be?

3. What are models used for?

Part I

ACTIVITY 1
MODEL BOAT DESIGN BRIEF

4. What is a system?

5. Have you ever designed a project or built something with tools?

If yes, describe your project and list your major process stages from concept to finish. If no, think about and list what steps you might go through to design and build a boat hull for higher performance.

6. Describe an experiment that determines which of two boats has greater acceleration.

1

Part I

SCI LINKS. THE WORLD'S A CLICK AWAY

Topic: modeling
Go to: www.scilinks.org
Code: CAB02

SCALE MODELING AT WORK

Testing Ship Performance

In shipbuilding, scale models are basically small versions of the ship that the designer wants to make. It is important that the model look like the ship and act like the ship. Ship designers often do their testing on models to develop cost-effective, high-performance ships. Testing is cheaper, less time-consuming, and safer when carried out on models rather than on full-scale ships.

By placing the model in a test tank, designers can measure handling characteristics such as stability in heavy seas, resistance to forward motion, and power required to move at a certain speed.

Sometimes ship designers must build and test several models, particularly if they are working under strict requirements regarding speed, fuel cost, and the ship's effect on the environment. Years ago, one or two tests were enough to ensure that the actual ship would satisfy design requirements. Today, however, ship designers must find ways to lower fuel costs and reduce pollution from engines, so it has become more important to design efficient hulls. Designers often test models and make modifications many times for each design.

One measure of ship designers' success with models is how much they can reduce the cost of running a ship over a 20-year lifespan. It is not unusual to obtain a 10–15% improvement as a result of careful model testing. These gains can be obtained by making small changes in the shape and finish of the hull, the size of the propellers, or the power of the engine.

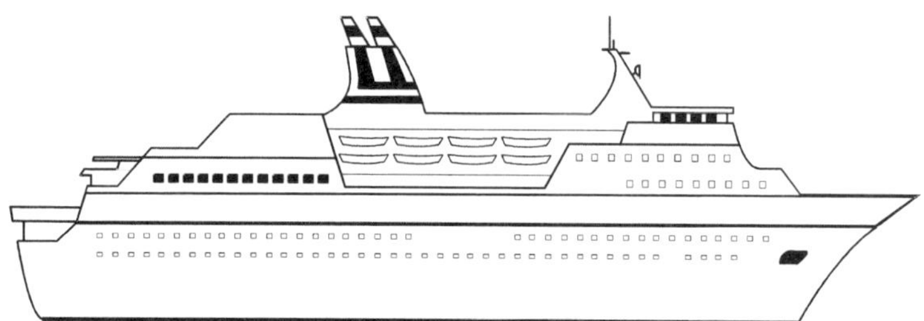

On new cruise liners, fuel costs have been reduced by 10–15% through model testing.

Part I

DAY 1

ACTIVITY 1 Teacher Pages

OVERVIEW: MODEL BOAT DESIGN BRIEF

Give students the Model Boat Design Brief and the Snapshot of Understanding. Have students read the design challenge. Working individually, students answer questions about their prior experience with modeling and with designing experiments. Initiate class discussion and highlight important design issues.

As an optional homework assignment, students might read about the history of boatbuilding to understand boat design as it has evolved from a craft to the highly developed computer modeling techniques of today.

Preparation

- Read and become familiar with the entire unit.
- Photocopy student activity sheets for distribution.
- Define your assessment system with a clear, simple description.
- Locate optional homework readings that feature local boatbuilding activities: traditional or contemporary.

Materials

- Student Activity Sheets
    - » Course Outline
    - » Overview: Model Boat Design Brief
    - » Snapshot of Understanding
    - » Scale of Modeling at Work
- Ring, Pocket, or Folio Binder (student supplied; for keeping student activity sheets, notes, laboratory journal, and drawings for reference and portfolio)

Time Requirement

This activity requires one class session. Completing the Snapshot of Understanding takes about 20 minutes.

1

Teaching Suggestions

Introduction

Hand out the Model Boat Design Brief student activity sheets. Ask students to keep these and future sheets together and to bring them to the classroom with other notes to serve as a record and reference for daily activity (and assessment) in the unit. Advise students that they will work in teams, using processes of technological design and scientific inquiry, and that other teams will critique their prototype with respect to the challenge criteria.

Students are also required to document their activity in a laboratory journal in order to contribute effectively to the final team presentation and to enhance their individual portfolios. Be clear on your rubrics for assessing their work and share them with students. Indicate which activities will be individually graded and which will be given a team score. Be prepared to justify team scoring if some students (or parents) are not used to the idea.

If computer modeling, computer-aided design (CAD), or computer-aided modeling (CAM) activities will be involved, you may want to mention this at the beginning of the unit, but hold introduction of the equipment until a later class.

Part I

ACTIVITY 1
TEACHER PAGES

Preassessment

Hand out the Snapshot of Understanding. Emphasize that it is not a test and that students will not be graded on this activity. The purpose of the Snapshot is self-diagnostic—to find out what students know initially about the key science and technology learning objectives of Construct-a-Boat.

An inventory of students' prior knowledge is an important teaching and learning tool. Not only does the inventory help guide students toward the concepts they need to learn the most, but it also prepares them to accept new information in a manner that ties meaningfully to what they already know. At the end of Construct-a-Boat, students will be able to compare answers given at the beginning of the unit with those they will answer at the end of the unit.

Collect and retain the Snapshots after students complete them.

Computer Use

This unit can be conducted with little or no computer use, or it can involve some or all aspects of computer modeling, data collection, design, and manufacturing. If you decide to use computer modeling, you will need to introduce these concepts briefly and indicate what level of expertise you expect students to acquire in these areas.

DAY 1

ACTIVITY 2
Quick-Build Model Boat

OVERVIEW: QUICK-BUILD MODEL BOAT

The quick-build is a simple model boat that you build to begin exploring the variables that affect boat performance. Your team can build it quickly, within one class period. Use the specified materials to build this model, and follow the technical instructions. You will then use the quick-build to make some baseline measurements. *Do not make any design changes before you make baseline measurements!*

Quick-Build Specifications

This section provides materials and tools lists, technical instructions, and technical drawings. It is important that you build according to the technical instructions and drawings provided. The quick-build ensures that every team starts in the same way, and you will be assessed on the *improvements* you will make to the quick-build's baseline performance. You must make the baseline measurements before you redesign anything so that you can measure improvements accurately. Be sure to check with your teacher before you make any substitutions for materials.

Materials and Tools for Each Team

Materials

- polystyrene (hull)
- 9-volt battery
- #22 AWG wire
- fan
- battery connector
- binder clip
- hot glue
- large paper clips
- electrical tape
- switch

Tools

- saw
- pliers
- wire strippers
- wire cutter
- sandpaper
- hot glue gun
- metric ruler

Part I

ACTIVITY 2
QUICK-BUILD MODEL BOAT

Technical Instructions

1. Cut the polystyrene to meet requirements for length and width of the scale model hull according to the blueprint.
2. Position the fan 2 cm from the quick-build stern. See the blueprint for details.
3. Build the switch according to the electrical switch detail drawing.
4. Build the electric system according to the electrical drawings.
5. Place the 9-volt battery and switch on the top surface of the hull. The waterline is where the surface of the water meets the hull. Position the battery and switch so that the waterline is parallel to the deck. Fasten them in place using glue or tape.
6. Straighten the two outer bends of each of two paper clips, and insert the straight end into the center of the hull. The paper clip will look like an upside-down "J." See the blueprint for details.
7. All lengths must be accurate within 2 cm. This type of acceptable error margin is called *tolerance*.

2

Blueprint

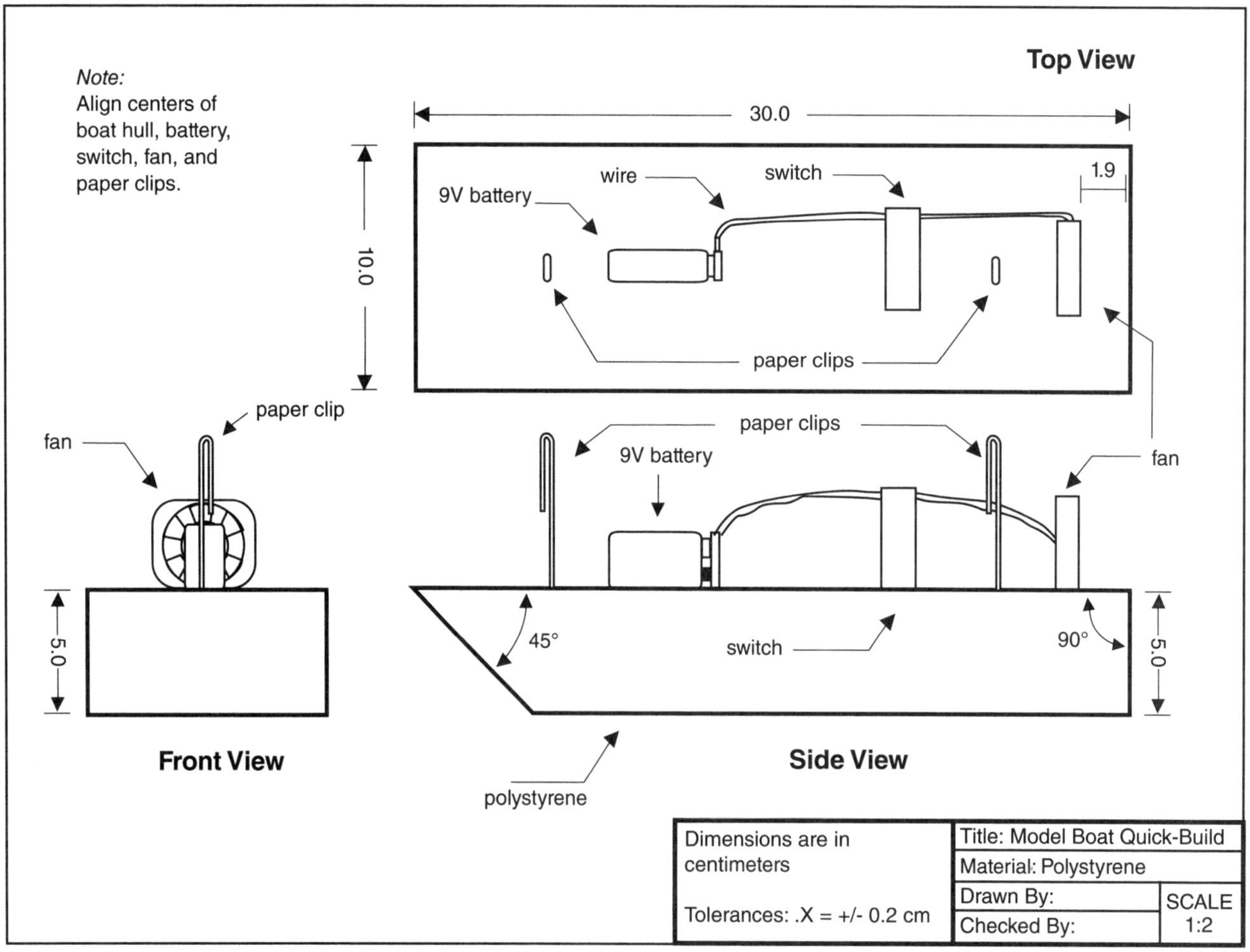
Note:
Align centers of boat hull, battery, switch, fan, and paper clips.
Top View
30.0
10.0
9V battery
wire
switch
1.9
paper clips
paper clip
fan
paper clips
9V battery
fan
45°
switch
90°
5.0
5.0
Front View
Side View
polystyrene
Dimensions are in centimeters
Tolerances: .X = +/- 0.2 cm
Title: Model Boat Quick-Build
Material: Polystyrene
Drawn By:
Checked By:
SCALE 1:2

Part I

ACTIVITY 2
QUICK-BUILD MODEL BOAT

Electrical Drawings

These drawings illustrate two different ways to show how the parts are connected.

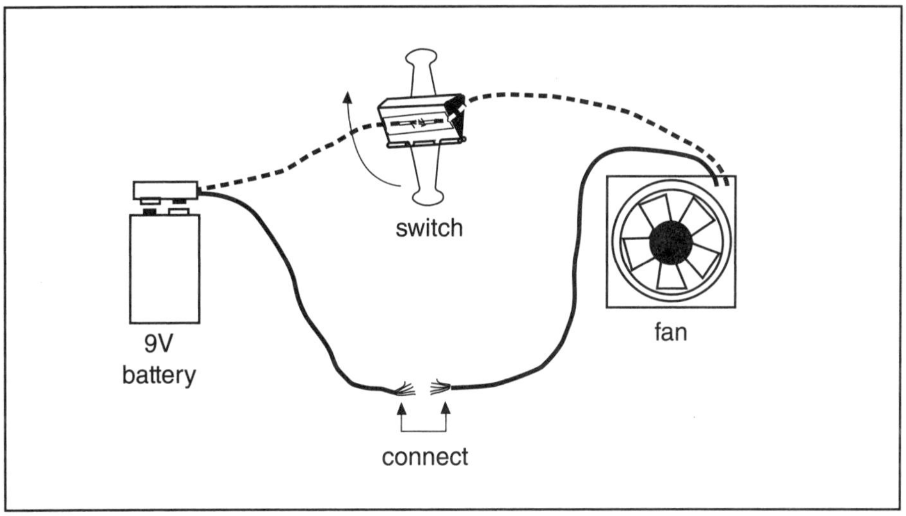

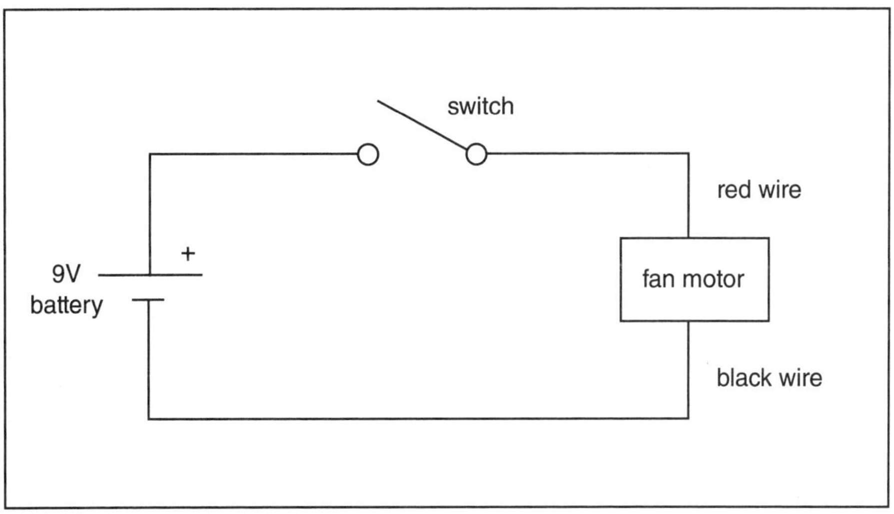

2

Electrical Switch Detail Drawing

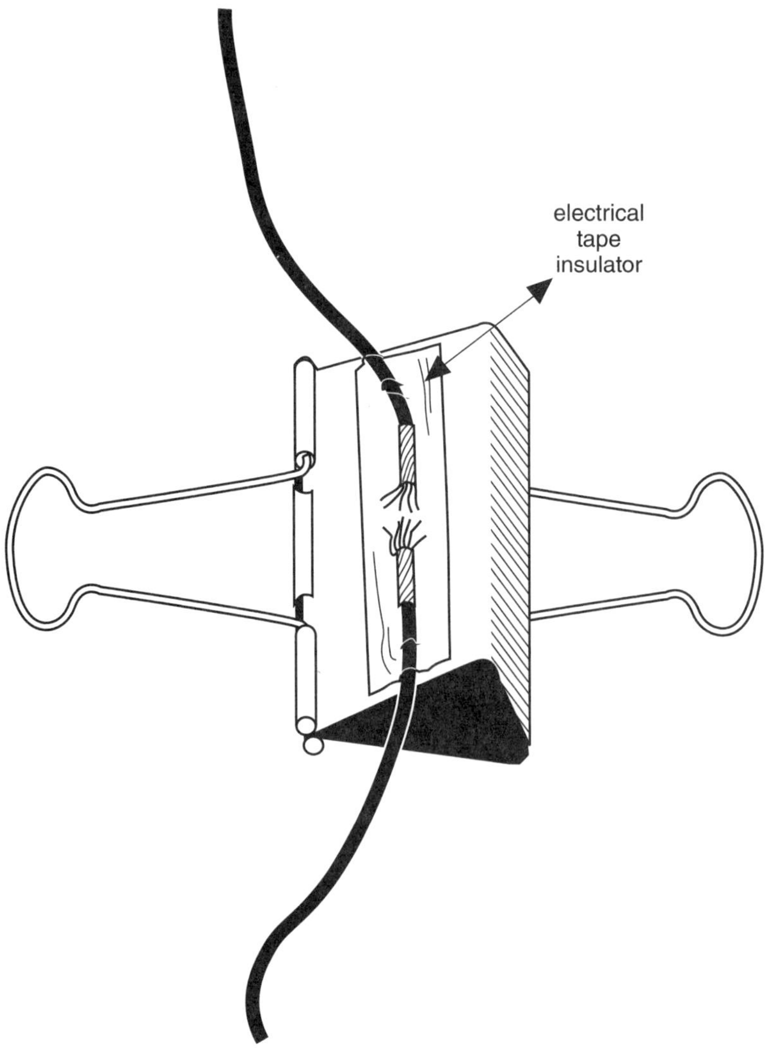

Part I

ACTIVITY 2
QUICK-BUILD MODEL BOAT

Making a Test Tank

Materials List

Part	Type	Quantity
Sides	wood 2.5 m × 5 cm × 10 cm	2
Seal	6 mm plastic sheet	1 roll, 60 cm wide
Corner bracket	corner	30 cm
Nails	8 d	32
Duct tape		1 roll
Guidepost		2
Guidewire	monofilament fishing line	1 roll

Technical Instructions

1. Fasten guideposts so that they are centered at each end of the tank.
2. Tie the monofilament fishing line to each guidepost, making sure that the height is adjusted so that the line can run through the approximate center of the paper clips on the quick-build. Your ability to do this will be affected by the height of the water.

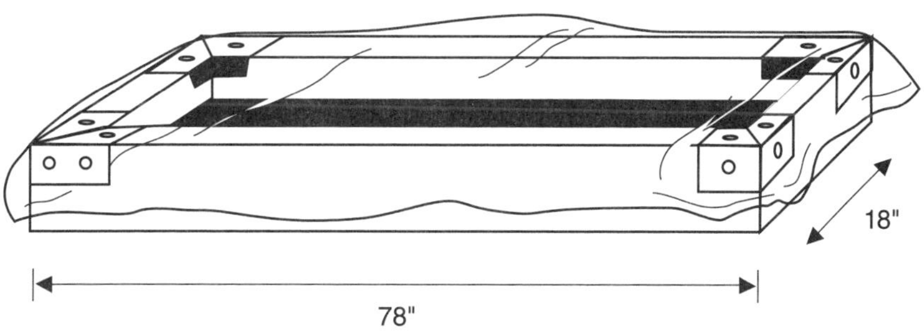

2

Exploring the Model Boat System

A system is a collection of interacting elements that function together as a unit. By studying the design of a boat as a system, you begin to understand why a boat performs well or poorly. Then you can improve the boat's performance by redesigning the most important elements of the system.

The purpose of this activity is to explore the performance of your quick-build before you make any design changes. As you explore, keep track of your observations and hypotheses about what affects the speed of the boat.

Observe

Run your quick-build along the water tank and observe what happens to the boat, the guidance system, the water, and other elements. You don't have to take any measurements, just make qualitative observations. Describe what you see and try to explain your observations in the space below.

Make an Educated Guess

Discuss with other members of your team which elements in the model boat system will affect its performance the most. Write down what you and your teammates think. Here are some questions to help you get started:

1. What makes the boat move? Is it the battery, the fan, the switch, the fishing line, or some combination of these elements?

ACTIVITY 2
QUICK-BUILD MODEL BOAT

2. What makes it hard for the boat to pick up speed?

3. You are challenged to redesign the boat hull. What part does the boat hull play in helping with the boat's performance? How does it affect the other elements?

4. Do you or your team have other ideas?

Systems Modeling Introduction Homework

A system is a collection of interacting elements that function together. Here are two examples of how to think about the boat steering control system.

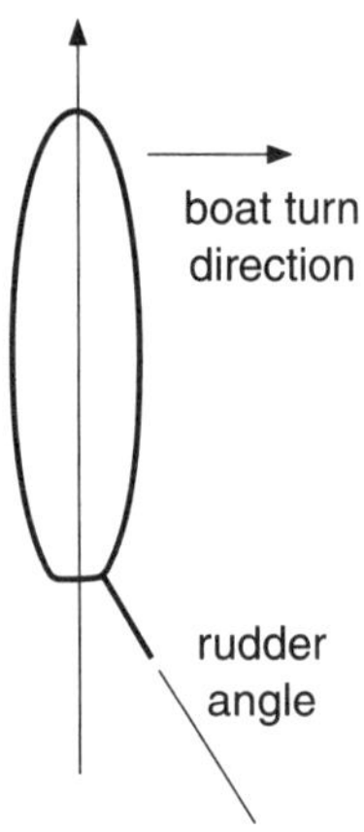

Example 1. The effect of the rudder on the boat direction. As the pilot increases the rudder angle, the turn rate increases. The *plus sign* means that an increase in rudder angle causes an increase in turn rate.

Example 2. The steering system of the boat, including the pilot.

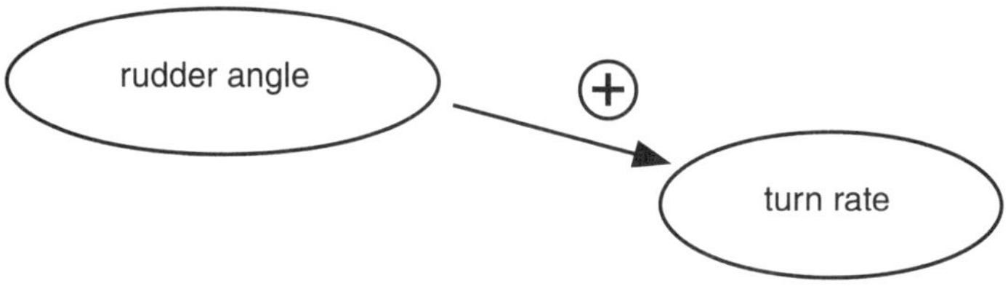

The boat pilot sees the error in the boat's direction and corrects by increasing the rudder angle. The *minus sign* means that an increase in rudder angle causes a decrease in the direction error. This is called *negative feedback* and results in a stable system—the boat stays oriented in the desired direction.

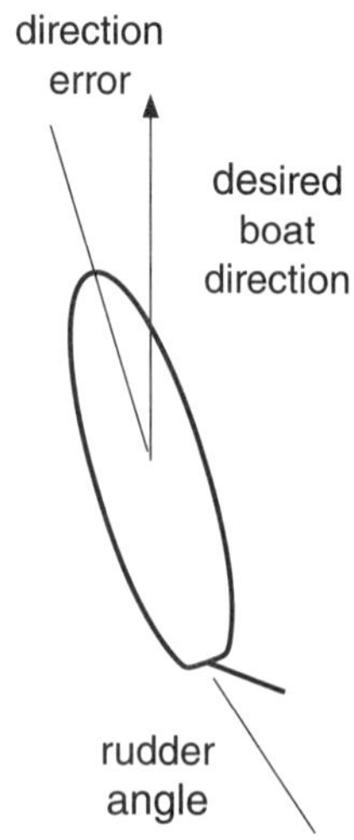

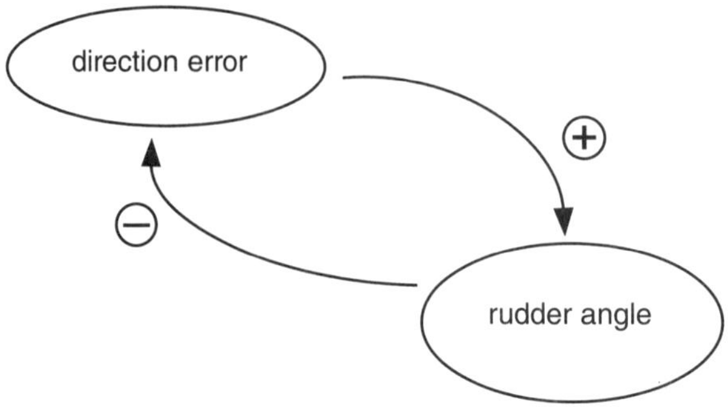

Part I

ACTIVITY 2
QUICK-BUILD MODEL BOAT

CAUSAL RELATIONSHIPS: LINKING VARIABLES

Evaluate the following statement and then answer the questions below: *An increase in speed causes an increase in water friction.*

1. Is this statement correct?

2. Why or why not?

SCILINKS® THE WORLD'S A CLICK AWAY

Topic: friction
Go to: *www.scilinks.org*
Code: CAB03

3. When a change in one quantity causes a change in another quantity *in the same direction*, we say that the first quantity has a *positive effect* on the second quantity. When a change in one quantity causes a change in another quantity *in the opposite direction*, we say that the first quantity has a *negative effect* on the second quantity.

In the diagrams below, show two distinctly different ways to complete the link, each containing (1) an arrow head and (2) a positive or negative sign.

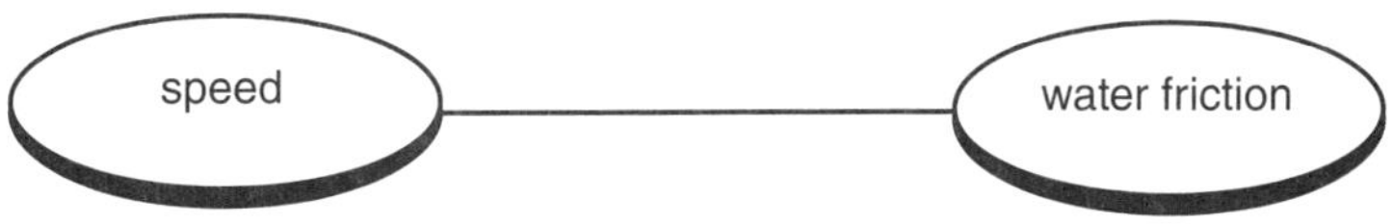

4. Combine the two links from step 3 to make one diagram.

5. State in words what your diagram shows.

Part I

DAY 1

ACTIVITY 2 Teacher Pages

OVERVIEW: QUICK-BUILD MODEL BOAT

Student teams make the quick-build according to technical instructions and drawings provided in the unit. Teams are challenged to exercise the communication skills of translating technical texts and drawings into a real object. They then explore the qualitative performance of the quick-build before making any design changes.

In a homework assignment, students are introduced to the use of systems diagrams to represent causal relationships.

Students are introduced to diagrams for linking functionally related variables. They explore the relation of speed to friction as it will impact their boat's performance.

Preparation

- Obtain quick-build construction and testing materials.
- Run through the quick-build assembly yourself.
- Modify instructions to fit your materials.
- Preview library and internet resources.
- Consider noise, safety, and access to water (access to AC power is optional) in choice of work site.
- Organize materials for orderly access.
- Determine the strategy for student team formation.
- Prepare a test tank.
- Contact local boatbuilders (optional).

2

Materials

For Each Student:

- Student Activity Sheets
    - » Overview: Quick-Build Model Boat
    - » Quick-Build Specifications
    - » Making a Test Tank
    - » Exploring the Model Boat System
    - » Systems Modeling Introduction Homework
    - » Causal Relationships: Linking Variables

For Each Team:

Materials

- polystyrene (hull)
- 9-volt battery
- #22 AWG wire
- fan
- battery connector
- binder clip
- hot glue
- large paper clips
- electrical tape
- switch

Tools

- saw
- pliers
- wire strippers
- wire cutter
- sandpaper
- hot glue gun
- metric ruler

Time Requirement

This activity requires one class session.

Part I

ACTIVITY 2 TEACHER PAGES

Teaching Suggestions

The quick-build class needs to be fast-paced. Due to the variation in fan motors and other elements of this lab, it is particularly important that you preview all aspects of the quick-build assembly before working with students. You should modify instructions to fit your circumstances. You should also feel certain that all teams will be able to complete the quick-build tasks within the class time assigned to each task.

Students will want to test their boats immediately, but you should delay performance measurements until the research section of this unit, where a framework for performance testing is developed. During the quick-build, emphasize frame questions to elicit qualitative answers such as "more" or "less." Questions requiring numerical answers should be postponed until later in the unit.

Tolerances and assembly drawings are included to provide a realistic machine-shop setting. If technology education is a high priority in your classroom, you will want to enforce strict adherence to this aspect of the unit. Otherwise you can decide which aspects best suit your objectives.

The quick-build and subsequent research sections stress systematic descriptions of qualitative relations. It is likely that these considerations will be new to your students. Do not be surprised if your students show disdain for the qualitative questions in these materials. Your students may also feel intimidated by the apparently simple, but in reality very difficult, reasoning that lies behind the qualitative questions. It is very important that you show your own enthusiasm for these questions and that you insist that students know how to handle them before moving on to quantitative calculations.

Research on the learning of mathematics and science has shown that standard modes of instruction place far too little emphasis on these essential aspects of mathematical thinking (Stigler and Hiebert 1999). Comparisons of expert and novice problem-solving practices show that a key characteristic of expert problem solvers is that they spend a great deal of time with qualitative issues before ever considering a numerical calculation. Novices tend to use a formula immediately, often without any understanding of its function or suitability.

Part I

3 DAYS

ACTIVITY 3 Research

OVERVIEW: RESEARCH

You have completed the quick-build and made rough observations of the Construct-a-Boat system. Your team's challenge is to improve the performance of your boat so that it will achieve a higher top speed as quickly as possible after starting. You will first need to establish some baseline measurements, then collect data on variables that are important to high performance, and investigate how these variables work with or against each other.

There are different ways to go about meeting your challenge. *Trial and error* is one method—fiddling around until something works. Trial and error is usually more effective when it is not an entirely random process; an educated "try" by an investigator who is alert to outcomes can yield shortcuts. However, time investment, cost-effectiveness of materials, and labor are usually important considerations that work against the benefits of trial and error.

A systematic approach to research involves careful planning and documentation, and this approach has several important strengths. Planning allows for division of labor, a sensible cost estimate, and a manageable time schedule. Estimates and schedules are essential if you want to finish the project on time and at a price you can afford.

Scope of Work

- Experiment with your quick-build to determine baseline data.
- Research the science concepts related to the performance of the boat.
    - » Identifying variables and effects
    - » Scaling
    - » Minimizing surface area
- Read science textbooks to study the relationships among force, speed, and acceleration.
- Explore the use of conceptual models, computer modeling, and scale models in explaining, designing, and evaluating a physical system.

Part I

ACTIVITY 3
RESEARCH

Baseline Measurements

Purpose

To show that you have met the challenge scientifically, you will need to compare your quick-build performance data with your final design. You can use the quick-build performance data as a baseline, and from that data you can tell if your design changes actually improve performance. Remember that your challenge is to redesign and build a boat hull that will (1) achieve the highest top speed and (2) achieve the greatest acceleration.

Collecting Performance Data

In the space below, write down a plan for collecting data to show (1) the top speed of a model boat and (2) the time it takes for the boat to achieve top speed. In your plan, describe what you are measuring, how you are measuring it, what calculations you need to make, how many people are needed to take measurements, and other pertinent information.

Process Alert!

You will probably revisit these research activities several times, because as you move on through the development activities, you will find that you need information you had not previously identified! No matter how carefully you plan, you may make mistakes; be alert to learn from them and adjust your plans.

3

Part I

SCILINKS THE WORLD'S A CLICK AWAY

Topic: presenting data
Go to: *www.scilinks.org*
Code: CAB04

Organizing the Data

Well-organized data will help you keep track of improvements in your designs and will help you make better designs in later stages. Carefully collect and record your data. Include notes on unexpected occurrences, such as someone bumping the water table. Organize your data so that you can make sense of it and use it later when you are redesigning your boat hull.

Set Up Tables for Recording Original and Calculated Data

Here is an example of recording and calculating data for the change of speed over time:

Time on Stopwatch (seconds)	Distance From Starting Point (meters)	Speed at Distance (m/s)	Speed Change = $\text{Speed}_n - \text{Speed}_{n-1}$ (m/s)

In your notebook or laboratory journal, set up tables for all variables that you have identified, measured, or calculated. Your tables should not look exactly like the one above, which is given only as a rough example. Below your table, leave space for notes on unexpected occurrences.

Part I

ACTIVITY 3
RESEARCH

Represent Your Data Using Graphs

Think about what you might want to study or show graphically. Graphs allow you to visualize relationships between the variables you are measuring and to see changes in your data. There are different kinds of variations over time: distance traveled, instantaneous speed, and acceleration all change over time. Here is an example of how you may represent this kind of variation:

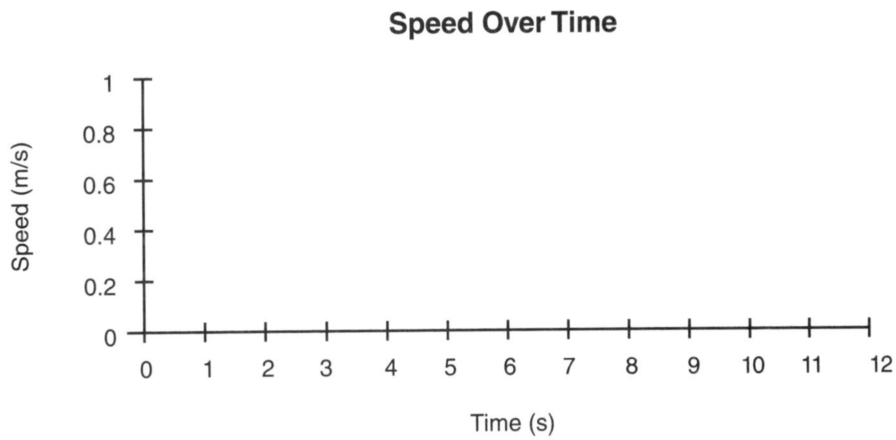

Make similar graphs for all variables that you have identified, measured, or calculated.

Are There Other Ways to Work With Data?

If you have access to a computer and spreadsheet software, such as Excel, try making the tables and graphs using a spreadsheet.

3

Identifying Variables

You have already made measurements of the speed of your quick-build, and you have probably noticed that the measurements vary over time. You will now identify the factors that could produce these variations.

Improved performance for the Construct-a-Boat challenge was defined as achieving higher top speed *and* reaching top speed in a shorter time. The scientific way to meet this challenge is to determine which variables affect improved performance. Once you've identified these variables, you can make sure to include these variables in your design changes.

Discuss variables with members in your work group, then complete the table below.

Variable	Why Is It Important?	How Do You Measure It?
Mass		
Force of the motor		
Friction of the water		
Wet surface area		
Surface roughness		
[add your own]		

Plan for Collecting Data for Model Boat Variables

Complete the table on page 36 for a plan to do your experiments. Make sure that your plan provides answers to these questions about *each variable* that you have identified:

ACTIVITY 3
RESEARCH

1. Is this variable directly measurable? If not, what can you measure to obtain quantitative data on this variable?
2. How do you do the measurement? Specify tools and comment on your expectation for accuracy.
3. Will you need to do any calculations from your measurements?
4. How do you calculate for the variable that you want to study? What assumptions or estimates do you need to make?

Use the space below to put down ideas, either your own or those of your teammates. These ideas do not have to be perfect or complete, and you do not have to write in complete sentences. Phrases, charts, or diagrams can sometimes do a better job, as long as you can come back to them later when you need to revisit your first thoughts.

SCILINKS®
THE WORLD'S A CLICK AWAY

Topic: mass
Go to: *www.scilinks.org*
Code: CAB05

Topic: force
Go to: *www.scilinks.org*
Code: CAB06

Measurement Plan

Variable	What to Measure	How to Measure	Calculations
Mass	Weight	Put model on triple beam balance	mass = weight/ (acceleration due to gravity)
Force of motor			
Friction of water			
Wet surface area			
Surface roughness			

Part I

ACTIVITY 3
RESEARCH

Working With the Variables

We want to know how each variable affects (1) the top speed of the boat and (2) the time it takes the boat to reach top speed.

Design a table (or some other way to record your data) so that you can work with the variables you have planned to investigate. Use the space on this page to record your ideas, and keep legible copies of your measurements for later reference.

Scale Model Preparation Homework

The shape of the scale model is the same as the actual boat you hope to design even though the model is smaller. You can calculate the linear scale factor by taking the ratio of the actual boat length to the model boat length. This will be useful to determine other dimensions of the model boat. Calculate what the linear scale factor would be for a 30 cm model of a 230 ft. boat and write it below.

Linear scale factor =__________

Use the boat design specifications and the linear scale factor to determine how wide the 30 cm model boat should be. The width of a boat is called the *beam*. For your ferry, the beam is 18 m.

Real boat length = ____________
Model boat length = ___________

Real boat beam = _____________
Model boat beam = ____________

Part I

ACTIVITY 3
RESEARCH

Applications of Scaling

Try using the linear scale factor to calculate the weight of the model boat. Explain why you cannot use the result. Use the space below to show your work and explanation.

Work out a strategy to estimate the weight scale factor for the model boat; it should allow you to calculate the weight in grams. Write the strategy below and include any assumptions you must make.

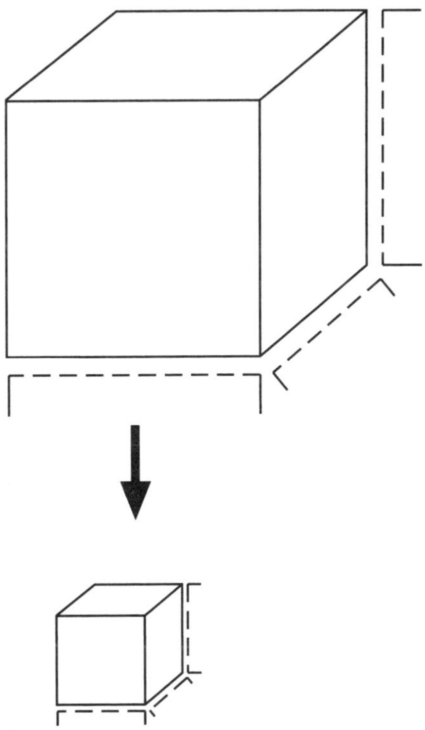

Weight scale factor = ______________

SCILINKS®
THE WORLD'S A CLICK AWAY

Topic: buoyancy
Go to: *www.scilinks.org*
Code: CAB07

Scale Model Extensions Homework

1. Use the weight scale factor to calculate the weight for the model boat based on the Design Brief data for a real boat with no cars or people. Write your solution below in grams.

Model boat weight = ____________

2. Estimate the maximum weight of people and cars that the ferry M/V *Nantucket* carries.

People and cars weight = ____________

3. Determine the scale weight for people and cars in grams.

People and cars scale weight = ____________

Part I

ACTIVITY 3
RESEARCH

Fluid Friction Dynamics

Boat designers and builders use the following mathematical model to determine efficient hull designs. The most efficient hulls are designed to have the least friction. The model is a formula:

$$\text{Friction Force} = C \times A \times V^2$$

friction coefficient (C) — wet surface area (A) — speed (V)

The *wet surface area* is the part of the boat hull that is in contact with the water. The *friction coefficient* is determined by the roughness of the wet surface area. The designer tries to create a more efficient hull by reducing the wet surface area. Builders can create more efficient hulls by reducing the friction coefficient. The trade-off for the builder is to balance the need for a low coefficient of friction with a durable, affordable hull surface.

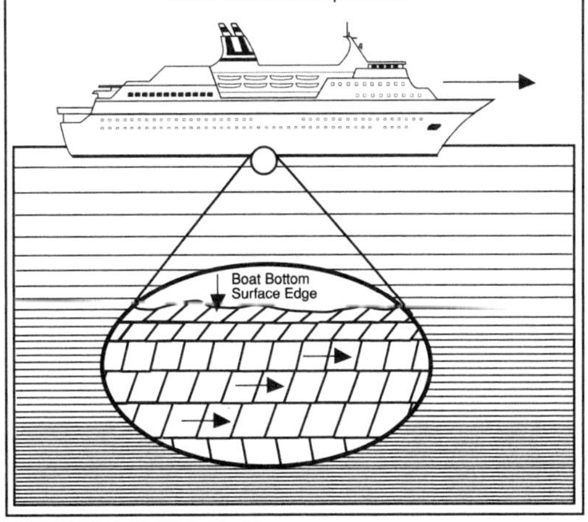

Fluid Friction Concept Model

In this fluid friction concept model, the ship slides through the water and pulls layers of water along with it. An enlarged view shows that layers closest to the boat hull are pulled the fastest. The roughness of the hull surface determines how much water is pulled along with the boat. This roughness is represented by the friction coefficient.

3

Minimizing Surface Area

The Question of Surface Area Versus Volume

Fluid friction force, which reduces the speed of the boat, is dependent on the friction coefficient and the wet surface area of the boat hull. One problem that a designer must solve is how to minimize that wet surface area. Given a specific volume, how would you choose a shape that will give you the smallest wet surface area of the boat hull?

What You Need

- measuring cylinder (1 l)
- spring balance
- cardboard three or more sheets 30 cm × 30 cm or longer
- glue
- markers
- fine, dry sand

Procedure

Using cardboard, build three rectangular open containers, with the sides and bottom to form the shape of a triangle, a square, and a semicircle. The top rectangular opening for all three boxes must be of identical measurements, and the height must also be the same for all three containers.

Record your measurements on the diagrams provided here.

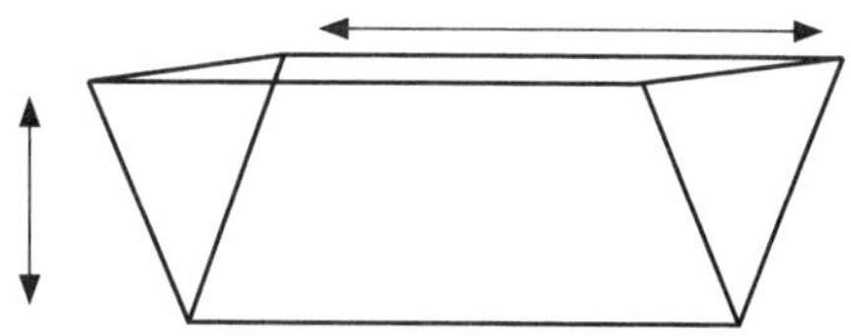

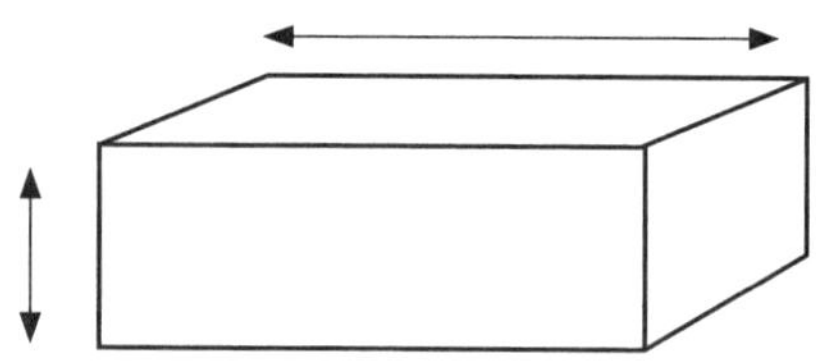

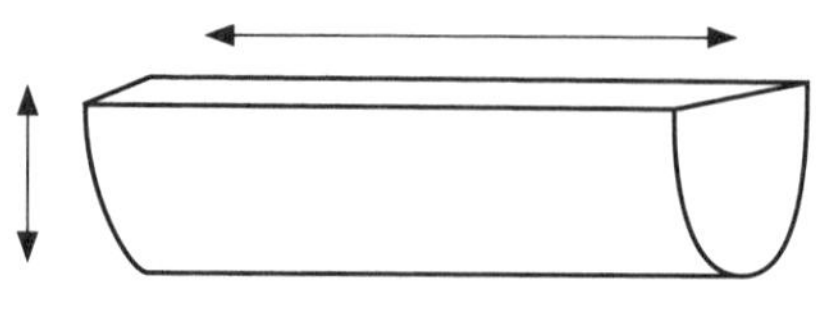

Part I

ACTIVITY 3
RESEARCH

1. Into each of the three containers, pour a carefully measured volume of 1,000 cc (1 l) of sand. Gently shake each one so that the sand is level when the container is standing flat on the tabletop. Mark the fill line all the way around on the inside surface of the container.

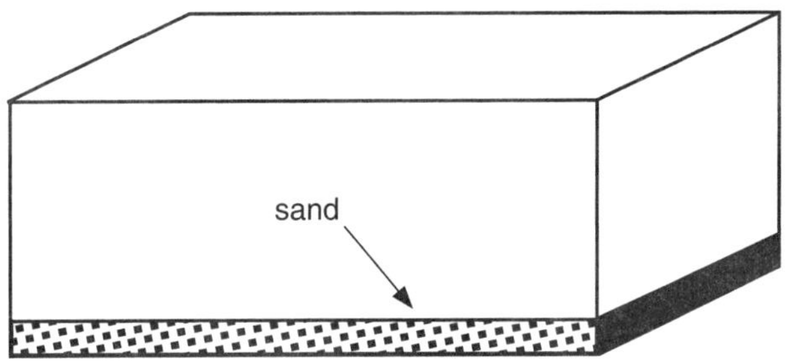

2. Discard the sand from each container and unfold each along the seams.

3. Compute the total area of the inside surface that touched the sand. Note the units you use for area.

4. What is your conclusion about the relationship between volume and surface area?

5. How will this information help you in the design of the boat hull?

Part I

3 DAYS

ACTIVITY 3 Teacher Pages

OVERVIEW: RESEARCH

Student teams work through a tightly structured series of activity sheets to familiarize themselves with the skills and structures they will need in the more open-ended development activities.

Students first make speed versus time measurements of their quick-build model to determine performance improvements through their design changes. The teams then list important variables and plan how to measure each variable.

As a homework assignment, students are introduced to scale modeling as an accurate three-dimensional (3-D) representation of the actual object, and determine some physical specifications for the model boat by scaling down from the real boat specifications in the Design Brief.

Using the Applications of Scaling activity sheet, students figure out why the linear scale factor does not work for weight (weight scales as volume, not length) and then determine a weight scale factor. They may use formulas from mathematics (length cubed) or empirically determine the scaling factor by cutting and weighing cubes of different sizes.

As a homework assignment, students use the weight scale factor to determine the model boat weight when empty and when loaded with people and cars.

In the Fluid Friction Dynamics activity sheet, students are introduced to the quantitative relationship linking speed and friction for a boat hull.

Finally, students complete the Minimizing Surface Area activity sheet. Using a constant volume of sand poured into 3-D shapes (open trays of triangular, rectangular, and circular cross sections), students explore the relationships between area and volume. This provides some baseline experience for designing a boat hull with minimum wet surface.

Note: This activity should be revisited after students do the Designer Problem activity sheet in the Development section. The initial introduction should be brief.

Part I

ACTIVITY 3
TEACHER PAGES

Preparation

- Review preparation issues for the quick-build.
- Prepare support materials for discussion of scaling.
- Define laboratory journal requirements.

Materials

For Each Student:

- Student Activity Sheets
    - » Overview: Research
    - » Baseline Measurements
    - » Organizing the Data
    - » Identifying Variables
    - » Scale Model Preparation Homework
    - » Applications of Scaling
    - » Scale Model Extensions Homework
    - » Fluid Friction Dynamics
    - » Minimizing Surface Area

For Each Team:

- quick-build boat
- stopwatch and meterstick
- graph paper
- set of 27 identical solid cubes and spheres of different diameters but identical material
- one liter of fine sand
- objects in a range of sizes and of uniform density

For the Class:

- accurate balance
- water tank
- computer-based or graphing calculator-linked position detector (optional)
- set of measuring cups (optional)

Time Requirement

This activity will take three class sessions: one for Overview, Baseline Measurements, and Identifying Variables; one for Applications of Scaling; and one for Fluid Friction Dynamics and Minimizing Surface Area.

Teaching Suggestions

The research portion of the Construct-a-Boat unit is tightly structured due to the relative complexity of the tasks required of the students and the need to ensure that all students acquire the skills necessary to advance to the development stage. If time permits, you may want to add library and internet search activities to this activity. A visit to a boatbuilder would also be an excellent extension option.

Devote the first class to the Overview, Baseline Measurements, and Identifying Variables sections. Most of the class will be conducted with teams working independently, but during final wrap-up, teams should share their ideas about variables to be measured and techniques of measurement. Each student should keep notes in a laboratory journal on both team and full-class activity.

The next class should be used for the exercise in applications of scaling, because this exercise may prove more complicated than it looks. For students who understand linear, quadratic, and cubic scaling, the exercise is simple (and will not require more than a few minutes). For students who do not understand scaling, this will be a difficult and highly important lesson.

The third class period is devoted to Fluid Friction Dynamics and Minimizing Surface Area. Most of this class will be taken up with experimentally investigating the relation of shape to surface area. An understanding of scaling is essential background for these measurements. If your students do not yet grasp scaling, you might delay this class until the middle of the Development activity (p. 50).

Part I

ACTIVITY 3
TEACHER PAGES

Day 1

Encourage students to develop their own methods and ideas through the Baseline Measurements and Identifying Variables activity sheets. You will need to observe their activities carefully. It is quite possible for students to design methods for measuring speed that do not work well in practice. Try to have the teams discover these errors themselves, perhaps by responding to your probing questions. If one team has developed a good system, ask that team to work with a team that is still struggling. Avoid giving all teams a set procedure because students will be able to implement a procedure without really understanding what they are doing.

If you have access to a computer or graphic calculator–linked position and velocity detector, you will want to set this up on the test tank. It may be necessary to add a large target sail-like structure on the boat to get good position data. You will need to pretest this setup and may have to modify quick-build instructions to ensure good data recording.

Use a brief, full-class discussion to make sure that all teams have a complete list of important variables. Students may wish to investigate variables that you feel are unimportant or are a diversion from the main focus of the unit. Try to find a suitable balance between cutting off their curiosity and allowing the class to be diverted to a completely different topic.

Day 2

If your class has difficulty with the material on scaling, use the optional second sheet on weight scaling. Provide each team of students with 27 identical cubes, which they can stack to create a large cube three units long on each edge. Also provide objects in a range of sizes and of uniform density. Have the students predict the weight of one size based only on the weight of a different size and a length measure from each. Allow students to repeat this prediction and measurement several times before attempting to explain why weight will be proportional to the cube of the linear measure.

Cubes and spheres are the best shapes for this exercise. Other shapes may introduce confusion concerning the best linear measure. If students have developed a fairly strong grasp of the concept, you may want to refine it by introducing cylinders of the same length but of different diameters. Here the weight scaling will depend on the square of the diameter (not the cube). It is also useful for students to see that cylinders of the same diameter but of different lengths will have weights that scale linearly.

Encourage students to explain these phenomena to each other, and be very patient with any difficulties students may have in understanding your explanations. Scaling is a very difficult concept to learn even though it is a simple and obvious concept to remember once it is understood.

One powerful way to stress the nature of cubic scaling is with two large cubes, the smaller being light enough to lift with some effort and the larger being impossible to budge.

Day 3

The student activity sheet Minimizing Surface Area assumes an understanding of the principle that a boat must displace a volume of water equal to its weight. Most students know of the Archimedes principle, but many might not really understand it. If this is the case, you could remind the students of the story of Archimedes in the bathtub.

To understand boats, it is important to distinguish floating objects from objects that are completely submerged. You might need to provide some time for informal investigation of these questions.

On the Minimizing Surface Area activity sheet, the volume of sand represents the volume of water that must be displaced by the vessel. The top of the sand represents where the waterline would be. Unfolding the container reveals the surface area that would lie below the waterline. Different shapes will yield different surface areas.

Some students might be troubled by the results of this investigation. The minimum surface area would be generated by a spherical-shape boat, yet no surface boats are designed as spheres.

We have been ignoring the effect of the frontal cross section. The need to design for a small frontal cross section is what forces designers to choose long thin designs of greater length than beam. The Minimizing Surface Area activity is useful in choosing between different possible cross sections. But minimizing surface area is not the only important consideration. The stability of the boat depends very much on the shape of

ACTIVITY 3
TEACHER PAGES

the cross section, and a good design must consider the trade-off between stability and minimal wet area.

Note

The mass versus weight issue is a tricky one, requiring teachers to give special attention to the actual measurement procedure and to students' misconceptions. *Weight* is the force due to gravity and is typically measured with a spring scale. *Mass* is the "amount of stuff" and is generally determined by comparing the sample to another standard sample. A triple beam balance is typically used to determine mass.

Confusion can arise if someone "weighs" something on a spring scale (as in this activity) and gives its weight in, say, grams, which is a measure of mass.

The relationship *weight = mass × force of gravity* allows us to indirectly measure a weight by measuring its mass on a triple beam balance.

4 DAYS

ACTIVITY 4 Development

OVERVIEW: DEVELOPMENT

You have completed activities investigating a model boat system. You have also researched the relationships among force, speed, and acceleration. To meet the Construct-a-Boat design challenge, your team will now develop a prototype that allows your redesigned model boat to achieve the highest top speed possible after starting in the fastest time possible. In designing and building this prototype, keep in mind that different variables may affect different aspects of the boat performance, and that you may not want to sacrifice too much of one aspect for the sake of another.

Scope of Work

- Redesign your boat hull so that your newly redesigned model boat will have improved performance over that of the quick-build.
- Build your prototype and test its performance.
- Collect performance data for the redesigned model boat, for comparison with the quick-build.
- Evaluate your design modifications.

Good planning is essential to good design. Look again at your research where you identified variables and wrote down key factors that affected performance. Think how best to combine design options to improve performance. Be creative within your objectives and constraints. Review the Inquiry Process (p. 69) and Design Process (p. 72) resource sheets (available from your teacher) for ideas on next steps. Evaluate your prototype critically and make modifications until you are satisfied with improved performance, or until you are simply out of time. Remember to reflect on your process, because how you go about your design, and what you learn, are key elements in the Communication activity to follow.

ACTIVITY 4
DEVELOPMENT

Designer Problem

Invent a new design that improves the performance of the model boat hull. Provide diagrams of the hull, including dimensions, and include explanations based on your research for how your design improves the model boat performance.

Here are some examples of hull shapes:

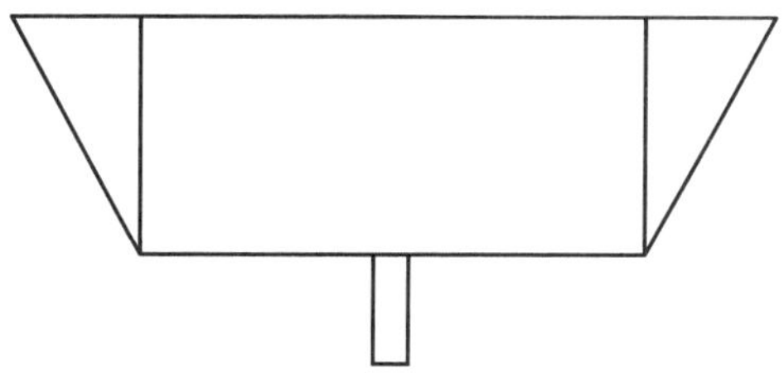

flat bottom

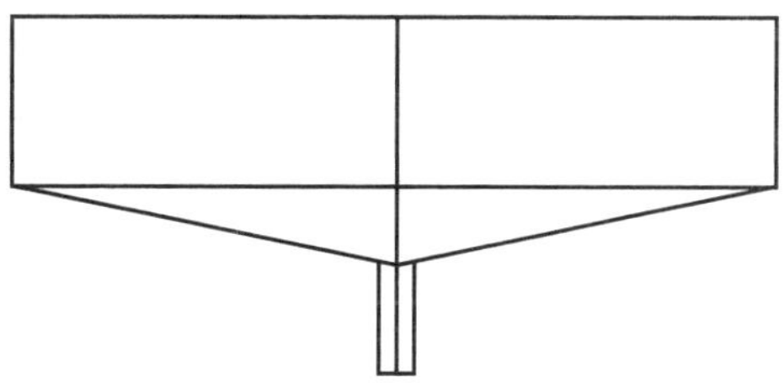

V bottom

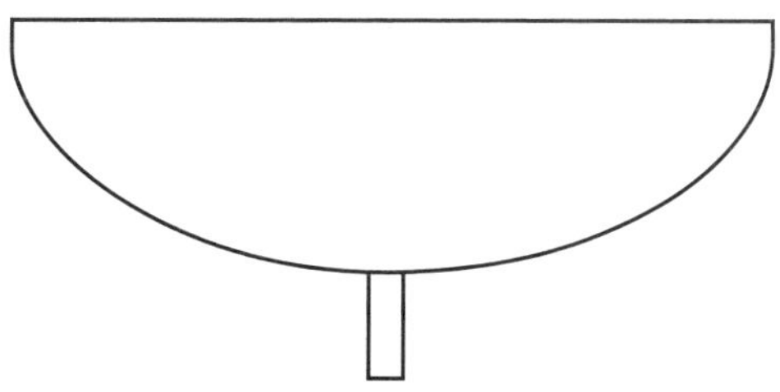

round bottom

4

Builder Problem

Propose *two* distinct methods to minimize fluid friction and describe them in the space below. Use additional pages for sketches if that helps your explanation.

Design and conduct a test to determine improvements in the friction coefficient for the model boat. If you can, try using the computer model to understand better how the friction coefficient affects the model boat performance. Describe your test below.

ACTIVITY 4 DEVELOPMENT

Prototype Construction

Proposal

Revisit the Design Brief and propose a design of a model boat that would show improved performance—a higher top speed and a shorter time to reach top speed. In the space below, summarize the features of your new design and the rationale for including those features.

Materials

List materials required for the construction of your redesigned model boat.

Schematics

On separate pages, make technical drawings for the construction of your new design. Include a side view, front view, back view, and top view. Specify dimensions in metric measurement units on each of the drawings.

Construction

Build your model boat according to the specifications of your team's design. If you must make modifications as you are building, take note of the modifications in your laboratory journal.

4

Evaluating Your Design

Purpose

Obtain performance data and compare these data with the baseline measurements, so you can tell whether your design changes actually improved performance.

Collecting Performance Data

Your challenge is to redesign and build a boat hull that will (1) achieve the highest top speed and (2) reach top speed as quickly as possible after starting. To show scientifically that you have met your challenge, you will need to compare performance data of the quick-build with those of your redesigned model. For the comparison to be valid, you need to collect data for your new model in the same way that you did for your quick-build. Collect data to show or calculate (1) the top speed of model boat and (2) the time it takes to achieve top speed.

Set Up Tables for Recording Original and Calculated Data

You have collected, recorded, and represented data obtained from working with your quick-build. Now collect data for your redesigned model boat. It is important to take your measurements in the same way that you did when you worked with your quick-build. You should also represent the new data in such a way that you can make unambiguous comparisons of the performance data sets. Pay special attention to the units of measurement.

Represent Your Data Using Graphs

When you make graphs from your data, try to work with a scale that accommodates the performance data for both the quick-build and the new design. Overlay the two graphs if you can.

Make similar graphs for all variables that you have identified, measured, or calculated.

Part I

ACTIVITY 4 DEVELOPMENT

Comparing Performance Data

1. Did your quick-build and/or the redesigned model reach a maximum speed?
   Yes / No
   top speed of quick-build =
   top speed of redesigned model =

2. How long did it take each of your two models to reach its maximum speed?
   acceleration time for quick-build =
   acceleration time for redesigned model =

3. Compare the performance of your redesigned model with that of your quick-build. Provide a brief qualitative description below.

4. How would you describe the improvement in performance quantitatively? What calculations do you need to make?

5. Summarize your results and explain how you have achieved them.

4 DAYS

ACTIVITY 4
Teacher Pages

OVERVIEW: DEVELOPMENT

The object of the work in this section is for students to redesign their boat hull to obtain improved performance over that of the quick-build.

In the Designer Problem activity, students consider alternative shapes for their hulls that reduce the wet surface area. Next, in the Builder Problem, students develop procedures to test for improvements in frictional drag. The Prototype Construction activity is an opportunity to put it all together. Finally, Evaluating Your Design lets students test their improved designs to compare boat performance data with the baseline measurements made with the quick-build.

Preparation

- Prepare tank test area.
- Arrange for team work areas.
- Prepare fire safety measures if polystyrene will be heat treated.
- Review laboratory journal requirements.

Materials

For Each Student:

- Student Activity Sheets
    - » Overview: Development
    - » Designer Problem
    - » Builder Problem
    - » Prototype Construction
    - » Evaluating Your Design
- Resource Sheets From Appendix A
    - » Inquiry Process
    - » Design Process

ACTIVITY 4 TEACHER PAGES

For Each Team:

Materials

- polystyrene (hull)
- 9-volt battery
- #22 AWG wire
- fan
- battery connector
- binder clip
- hot glue
- large paper clips
- electrical tape
- switch

Tools

- saw
- pliers
- wire strippers
- wire cutter
- sandpaper
- hot glue gun
- electric iron (optional)

Time Requirement

This activity will take a minimum of four class sessions. When possible, several days should be allotted to prototype construction, allowing students to undertake several redesign cycles.

Teaching Suggestions

Designer Problem

The quality of the Designer Problem activity depends strongly on the fabrication techniques you are willing to explore. In the simplest case, where only straight cuts are permitted, all designs will be limited to flat, angled surfaces. Curved surfaces are possible if you allow sanding (although this is difficult with polystyrene). Polystyrene may be melted with a hot iron and molded into virtually any shape. However, this technique does release fumes and may create a small fire risk. Therefore, it is not recommended unless you have access to a shop with appropriate venting and safety features. Working with wood instead of polystyrene is also an option if you have access to appropriate tools and work space.

Be sure students understand whatever limits you have imposed on the fabrication strategies before they begin the Designer Problem activity.

Builder Problem

There are two variables to consider in the Builder Problem activity: surface area and surface finish. Because the previous activity dealt with surface area, students will be largely concerned with surface finish for this activity. Fabrication techniques are very important for determining the range of available options. The measurement of surface drag has been left for students to define. They may wish to measure surface smoothness by feel or by measuring the force needed to drag some object across it. Alternatively, they may choose to measure the drag of the entire boat hull, a measurement that combines surface smoothness and surface area. If you feel this problem is too open-ended for your students, you may want to define the issue more precisely.

Part I

ACTIVITY 4
TEACHER PAGES

Prototype Construction and Evaluating Design

This last exercise should pull together the concepts covered in all preceding activities. On the one hand, student teams should be given as much freedom as possible to design and build the best boat they can. On the other hand, students should understand that they need to be using the knowledge gained from previous activities and not attempt to improve performance based on completely new variables (such as a large motor). Laboratory journals should be referenced on previous activities and ideas.

3 DAYS

ACTIVITY 5 Communication

OVERVIEW: COMMUNICATION

You have met the challenge. You have redesigned a boat hull and you have built and tested a model boat that has improved performance over the quick-build. It has taken you some time and much effort, but until you communicate the importance of your work to the people who matter, your efforts will have little meaning. Indeed, one of the most critical abilities today is to be able to communicate clearly, effectively, and persuasively.

As an engineer, scientist, or developer, you are dependent on funding from private and public foundations. It is important to convince those with money why they should give support to the work you do. You may want to present your ideas to local officials or create a web page that would make your work accessible to the entire world. There are individuals and companies in the boatbuilding industry who are very interested in new designs or marketing their boats to consumers. Publications are also important for communicating your findings to the greater scientific community.

Think about the discoveries you have made with the Construct-a-Boat challenge. Communicate the important parts of what you have done to several different audiences, including your classmates, a group of novice boatbuilders, and, most importantly, yourself. Think about the interests your audience might have in your work. Recognize that there are many possible formats to communicate with your audience, and select one or more formats to present your work.

Scope of Work

- Create a project report that communicates the results of your work. Your target audience may be executives of a boatbuilding company or people attending a trade show.
- Present your redesigned model boat, including the rationale, substance, and outcomes of your effort.
- Complete the postchallenge assessment and reflect on what you have learned in this unit.

Part I

ACTIVITY 5
COMMUNICATION

Creating a Project Report

The final activity of this unit is for you to communicate the results of your work in a project report. Your goal is to include information that will inform builders of full-size boats about the capabilities of your design.

An important skill for you to demonstrate in your project report is the ability to communicate clearly in writing. The writing of each section should be well organized and clear enough for someone unfamiliar with your design to understand.

Have fun thinking about ways to creatively describe your model design. Consult the table of contents below for a list of the topics you should address in your project report. Members of your team should divide the responsibility for each of the sections that need to be written.

I. Project Statement

II. System Overview
   A. Physical Specifications
   B. Performance Specifications

III. Research

IV. Development

V. Supporting Data

5

Presentation

In meeting the Construct-a-Boat challenge, you redesigned a boat hull and built a model boat for improved performance. You have kept a careful record (perhaps even photographs), and you have the prototype model for display. You are now asked to give an oral presentation to an objective audience on the rationale, substance, and outcomes of your effort. You have limited time for your presentation, so do some careful planning and rehearsal. This should be a team effort, with each team member responsible for communicating a key part of the presentation. Expect to field questions from your teacher and audience. You may find visual aids useful in presenting key data and in providing your audience with tools for quick and clear analyses.

Prepare an outline of key points to cover. Focus on capturing your audience's interest, but clearly identify the strengths and distinguishing features of your design as compared with those in other team presentations. Present the evidence that the design accomplished its purpose by quantitative and qualitative comparisons with the quick-build, using the criteria developed previously by the class.

List your preliminary presentation outline with team member assignments below. Consult with your teacher as a resource in your planning.

Presentation Outline	Team Member Assigned
Title:	
I.	
II.	
III.	
IV.	

ACTIVITY 5
COMMUNICATION

Reflection and Recommendations

Set forth at least five prioritized factors to consider, design criteria, or statements of wisdom you would offer to others who would be interested in designing and building a high-performance boat. Make your recommendations based on the experience and knowledge you have gained through research, development, and testing of your own model boat, and through comparing your results with those of other teams.

Using a model of causal relationships, describe two key factors you enhanced to improve performance.

What do you recommend about working with performance evaluation before beginning your important design work?

5

SNAPSHOT OF UNDERSTANDING
What I Now Know About Models, Systems, and Design

1. What are some of the factors that would have significant effects on the speed and acceleration of a model boat?

2. Do these factors have any effect on each other? If so, what might those effects be?

3. What are models used for?

ACTIVITY 5
COMMUNICATION

4. What is a system?

5. Describe your project and list your major process stages from concept to finish.

6. Describe an experiment that determines which of two boats has greater acceleration.

Part I

3 DAYS

ACTIVITY 5 Teacher Pages

OVERVIEW: COMMUNICATION

Student teams summarize their learning by creating a product prospectus that includes information about specific applications for their model boat, its construction, hull shape, factors influencing the performance, and performance parameters.

In the final assessment, students answer questions similar to those at the very beginning of the unit in the Snapshot of Understanding.

Preparation

- Prepare a presentation event plan, location, and agenda.
- Consider offering spectator invitations.
- Provide students with examples of project reports.
- Arrange for use of word processing/graphics (or CAD) computer stations.
- Prepare a grading plan for your evaluation of the team and individual effort.
- Customize the project report table of contents to fit the time and educational objectives for your class.

Materials

For Each Student

- Student Activity Sheets
  - » Overview: Communication
  - » Creating a Project Report
  - » Presentation
  - » Reflection and Recommendations
  - » Snapshot of Understanding

For Each Team

- Select presentation aids from the following options:
  - » overhead projector
  - » flip charts
  - » computer displays

Part I

ACTIVITY 5
TEACHER PAGES

Time Requirement

Devote one or more classes to team preparation of the reports. Presentation of the reports should be strictly timed so that all teams can report within the allotted time (one or two days, depending on class size).

Devote one class to reflection on the process and final self-assessment. Allow about 20 minutes for students to complete the Snapshot of Understanding.

Teaching Suggestions

Preparing the Report

Teams should divide preparation and presentation responsibilities among members so that effort and benefits of the work and learning are distributed equitably. An assessment rubric tailored to fit your class learning and evaluation objectives should be made available to teams well in advance—possibly at the beginning of the prototype development phase—to encourage maximum awareness, planning, and preparation.

Presentation

Fitting a presentation into the time allotted is a skill few speakers ever master. You will need to warn all teams of the importance of staying within their time slot, and you will need to be strict in enforcing time limits. If you invite visitors to the presentations, they will need to be informed of your time limits and of the kind of input you would like them to provide.

Completing the Snapshot of Understanding

After students complete the final Snapshot of Understanding, provide a brief period of time for them to compare their new answers with those on their preunit Snapshot.

APPENDIX A
Side Roads

The material in this appendix is intended to support activities that you may choose to add to those described in the core unit. Many of these are key activities, but they have been placed in this appendix because they can fit in several different places in Construct-a-Boat—exactly where they are used is a matter best decided by you in response to student questions and feedback. Some activities may be profitably used more than once. An analysis of the design process, for example, will provide different insights when used in the research activities than when used in the development activities.

Two optional Snapshots of Understanding are included. One can be used to evaluate students' grasp of the concept of a model; the other to evaluate student understanding of control of variables. Construct-a-Boat assumes that students understand these concepts. You will need additional materials if students' understanding of these concepts proves to be weak.

In this appendix:

- Inquiry Process
- Modeling Design Solutions
- Design Process
- Charting a Mathematical Model
- Simulating Model Calculations
- Snapshot of Understanding: Models
- Snapshot of Understanding: Control of Variables

APPENDIX A
SIDE ROADS

Inquiry Process

The inquiry process is often viewed as a cycle of action that repeats until the investigators reach a satisfying solution. It can be described with seven basic elements:

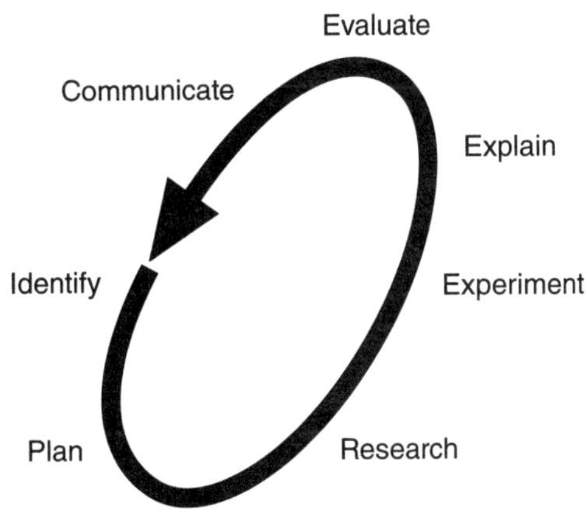

- *Identify* and clarify questions. Understand the issue or problem, and make a testable hypothesis.
- *Plan* appropriate procedures. Brainstorm, draw and write ideas, clarify their ideas, and suggest possible strategies or methods.
- *Research* major concepts. Learn what is known about the situation from sources other than actual investigation, and obtain information from preliminary experiments. Decide what technology, approach, equipment, and safety precautions are useful. Document your experiments and log your data.
- *Experiment*. Use tools and measuring devices to conduct experiments. Use calculators and computers to store and present data.
- *Explain* logical connections. Analyze your data. Formulate explanations using logic and evidence, and possibly by constructing a physical, conceptual, or mathematical model.
- *Evaluate* alternatives. Compare your explanations with current scientific understanding and other plausible models. Identify what needs to be revised, and find the preferred solution.
- *Communicate* new knowledge and methods. Communicate results of your inquiry to your peers and others in the community. Construct a reasoned argument through writing, drawings, and oral presentations. Respond appropriately to critical comments.

SCILINKS® THE WORLD'S A CLICK AWAY

Topic: scientific inquiry
Go to: www.scilinks.org
Code: CAB08

Questions

Read the following questions, but do not answer them until after your team has experienced working together on the design challenge research activities.

1. Make your own checklist of team activities that correspond to steps in the cycle described on page 69:

2. Create your own version of the inquiry process using words and pathways that fit your team's activity.

3. What shape is your inquiry pathway diagram (circle, spiral, cascade, other)?

4. How and where do the seven steps described above fit within your process description?

APPENDIX A
SIDE ROADS

Modeling Design Solutions

Modeling is the activity of imitating reality. You can use modeling to simulate actual events, structures, or expected conditions to test, analyze, and refine your design ideas. You also use models to focus on important parts of the total problem.

In Construct-a-Boat, we use three different types of models:

- *Graphic models:* Typical graphic models are conceptual drawings, graphs, charts, and diagrams. Conceptual drawings capture the designer's ideas of specific details; graphs and charts display numerical information and help the designer assess results; and schematic diagrams show relationships between components. You have already made and used several graphic models.

- *Physical models:* A physical model is a three-dimensional representation of an actual object. You can construct a physical model with materials that are easy to work with, such as wood, clay, polystyrene, and paper. Because full-size models are often impractical, people use *scale models* to show how a product will look or to test the operation of a system. A scale model is proportional to actual size by a ratio. Your quick-build is a physical model.

- *Mathematical models:* Mathematical models show relationships in terms of formulas. For example, the formula for speed shows the relationship between the distance traveled by a boat and the time it takes to travel that distance:

$$\text{Speed} = \frac{\text{Distance}}{\text{Time}}$$

  In general, you would need to use many formulas and interrelate them to predict the results of more complex relationships.

A

Design Process

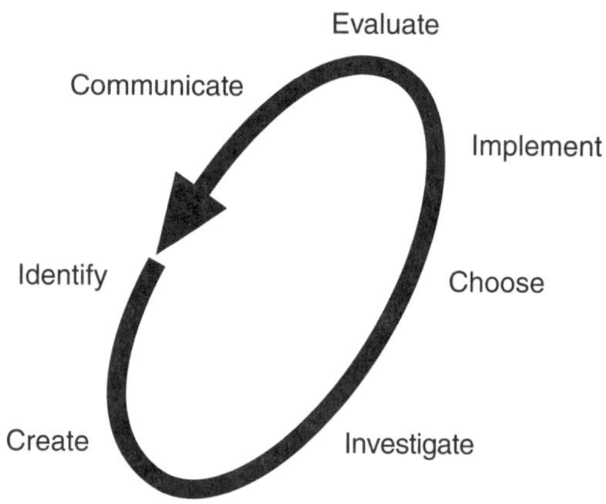

The design process is often viewed as a cycle of action that repeats until the designers reach a satisfying solution. It can be described with seven basic elements: identify, create, investigate, choose, implement, evaluate, and communicate.

- *Identify* and clarify the situation. Understand the challenge or problem, including the criteria for success and constraints on the design.
- *Create* solutions. Brainstorm, draw and write ideas, and suggest possible strategies or methods.
- *Investigate* possibilities. Learn what is known about the situation, and what technology or approach could be useful. Conduct experiments to test your ideas.
- *Choose* a solution. List the solutions most likely to be successful, and make decisions for how well each solution meets the design challenge or solves the problem.
- *Implement* the design. Learn that a successful design often depends on good fabrication, whether it is a scaled or life-size version of the product.
- *Evaluate* the design. Perform tests to obtain the feedback that informs them about the parts of the design that worked or needed improvement.
- *Communicate* the solution. Present your designs to your peers and others in the community, communicating your ideas through drawings, writing, formal presentations, informal discussions.

APPENDIX A
SIDE ROADS

Questions

After reading about the design process, answer the following questions:

1. Which elements of the process have you already experienced?

2. Which elements have you not yet experienced?

3. Where in the process do you think you are now?

4. What will your next steps be?

Charting a Mathematical Model

In order to study how various factors influence your boat's speed and acceleration, you need a way to look at several factors acting at once. A chart can be a useful way to do this. Study the chart below and fill in the incomplete cells.

Factor	What Does It Depend On?	What Does It Affect?	How Would You Control It?
Mass	Building material, size of boat, load being carried	Acceleration (speed of change)	Choose different kinds of material, work on size of boat
Force of motor (thrust)	Batteries, type of propellor, size of motor	Acceleration	
Force of friction (drag)			
Acceleration (speed of change)			

Part I

APPENDIX A
SIDE ROADS

Quantifying the Chart

With your factors now in place and with a qualitative feel for how they relate to each other, you can start to see how they affect each other quantitatively. Because the boat starts with no speed and must accelerate up to its maximum speed, you need to examine acceleration. The fundamental law for acceleration is Newton's second law: $F = ma$.

First, look carefully at the three components of this equation. There are two main forces acting on our boat: the force of the motor and the resistance or friction force of the water. There is one mass, the mass of the boat, and one acceleration, the acceleration of the boat.

An equation for friction force is $F_{friction} = kv^2$ where k is made up of two parts: C, a measure of roughness; and A, the total surface area in the water. Because neither C nor A changes as the boat moves through the water, consider their product k as a constant number. (In hydroplanes, A does change as the boat moves, but you are not working with that kind of boat.)

The equation for $F_{friction}$ has the value zero when the speed is zero. You can now fill out the first column of the chart on the next page.

Use the letter M to stand for the mass of the boat; you will need to weigh yours to get a number. Enter that mass all across the row because it will not change.

Use the letter F to stand for the force of the motor (see unit of measuring the force of the motor). Enter that all across the row because it will not change.

A

Time (seconds)	0	1	2	3	4
Mass	___ g	___ g	___ g	___ g	___ g
Force of Motor (thrust)	*F*	*F*	*F*	*F*	*F*
Force of Friction (drag)	0				
Total Force	*F*				
Acceleration	*F/M*				
Speed	0				

Part 1

APPENDIX A
SIDE ROADS

Simulating Model Calculations

To fill in the second column of the chart on page 76, it may be helpful to work in teams of five. Each person takes a turn at calculating a value for the table. We will work up from the bottom row. Remember that the values you are calculating are relationships in a mathematical model, not necessarily numerical values.

1. The first person determines, from team data, the speed change value:

$$\frac{(\text{thrust} - \text{drag}) \times \text{time interval}}{\text{mass}}$$

2. The second person calculates the new speed:

$$\text{new speed} = \text{old speed} + \text{speed change}$$

   Enter this result for the next value of speed.

3. The third person, using team data again and the calculations above, determines the new acceleration value:

$$\frac{(\text{total force})}{\text{mass}}$$

   Enter this result for the new value of acceleration.

4. The fourth person calculates the friction of the water:

$$\text{drag} = \text{friction coefficient} \times \text{wet surface area} \times (\text{speed})^2 = k\,(\text{speed})^2$$

   Enter this result for the next value of force of friction.

5. The fifth person calculates the total force:

(thrust – drag)

Enter this result for the new value of total force.

Once the second column is filled in, move on to the third column. The first person repeats the calculation using numbers from the second column. Your team can continue this process until the boat reaches maximum speed.

Checking Your Intuition

You have completed one worksheet, and the data should help you see how design changes might affect the performance of the boat. What would happen if you made the boat lighter?

Use your intuition to complete the following sentences with one of these choices: *increases, decreases,* or *does not change.*

1. As the mass of the boat increases, the top speed ...
2. As the mass of the boat decreases, the top speed ...
3. As the motor force increases, the top speed ...
4. As the motor force decreases, the top speed ...
5. As the wet surface area increases, the top speed ...
6. As the wet surface area decreases, the top speed ...
7. As the gliding surface roughness increases, the top speed ...
8. As the gliding surface roughness decreases, the top speed ...

APPENDIX A
SIDE ROADS

Predicting the Effects of Design Changes

Using the chart calculation, you can see the effect of possible changes you might make to your boat's design. You can actually determine how a certain change will affect the time that it takes to reach top speed. If you know how to use a spreadsheet program, you can use it do the work of the five people and create a chart that fills in automatically.

Here are some design changes to consider:

- What happens when you make the gliding surface half as rough? How would you use such information to help you redesign and build your boat hull?
- How big a change would you need to make in the wet surface area to get as big an effect as you got cutting the roughness in half?
- How much would you need to change the mass of the boat to get this effect?

A

SNAPSHOT OF UNDERSTANDING
Models

Give three different examples of models.

What are models used for?

What is important to include in a model?

Can there be more than one model for the same thing?

APPENDIX A
SIDE ROADS

SNAPSHOT OF UNDERSTANDING
Control Of Variables

Two companies are conducting experiments to understand how to build a better canal system. Each company built two canals to learn how canals affect boat speed.

The first company built one canal that was narrow and shallow and another canal that was wide and deep. The second company built one canal that was narrow and deep and another canal that was wide and deep.

Which company designed the better experiment? Why?

APPENDIX B
Text Reconstruction Exercises

Forces, Speed, and Acceleration

The following paragraphs describe important information about forces, speed, and acceleration. To keep this information confidential, some of the sentences within each paragraph have been reordered. Your task is to restore them to their proper order.

¶ 1

___ Despite impressive mathematical accomplishments, the Greeks never understood the connections.

___ The relationship of force, speed, and acceleration eluded scientists for most of recorded history.

_4_ Even then, it took Newton over 20 years to develop his three simple laws.

___ It was not until Newton's efforts less than 350 years ago that a perspective consistent with our modern view emerged.

¶ 2

___ How can something as simple as Newton's laws be so confusing?

___ We therefore conclude that speed and force are related.

_2_ The confusion stems from our common experience of pushing and pulling.

___ In that experience, it takes force to make things move, and more force to make them move fast than to make them move slowly.

¶ 3

___ Here is a way to think about it.

___ But in Newton's view, speed and force are not related!

___ In Newton's famous second law, force is related to acceleration: the *change* in speed.

_3_ But why isn't the change in speed related to speed?

APPENDIX B
TEXT RECONSTRUCTION EXERCISES

___ How much you eat for dinner tonight is related to how much your weight will increase or decrease. ¶ 3 cont.

___ But even if I was told what you ate for dinner every night from the day you were born, I could not figure out how heavy you are.

___ Yet if I get on my bicycle, I know this is wrong. ¶ 4

___ Thus in Newton's picture of the world, force tells us about acceleration—the change in speed—but it cannot tell us about speed itself.

___ I have to pedal harder to go faster.

___ But the reason I have to pedal harder is because of air resistance.

___ Instead of pedaling harder, I could reduce my air resistance by getting racing clothes, or by cycling behind an air shield.

_6_ Then I would go fast without pedaling harder.

___ Thus a cyclist can go faster without increasing the force exerted.

B

Systems

The paragraphs below describe important information about systems. Your task is to restore these jumbled paragraphs to their proper order.

¶ 1

___ Understanding those effects and relations is called *systems analysis.*

___ Systems analysis and systems thinking are increasingly important aspects of the modern world.

_2_ In any complex situation, there are many related parts that affect one another.

¶ 2

___ The simplest kind of systems analysis involves understanding whether an increase in one quantity will cause an increase, decrease, or no change in another quantity.

___ If you increase what you eat for dinner, you will increase your weight (unless you also exercise more, or do something equivalent).

¶ 3

___ Often the links in a systems analysis can be complicated and even contradictory.

___ But when the weather gets colder, it also snows more and so you go skiing and get more exercise.

_2_ If the weather gets colder, you play less baseball and get less exercise.

¶ 4

___ In a simple case, we can tell what will happen without having to calculate with numbers.

_2_ It snows and I will go skiing.

APPENDIX B
TEXT RECONSTRUCTION EXERCISES

____ To calculate the exact amount of exercise for each case, I will need to use a mathematical model. ¶ 4 cont.

____ But will I get more exercise than I got back in July?

____ Here I will need to know how many times I played baseball, how many times I went skiing, and how much exercise I got each time.

B

Modeling

The paragraphs below describe important information about modeling. Your task is to restore these jumbled paragraphs to their proper order.

¶1

___ Our case is just the opposite.

___ The word *model* has several different meanings.

___ We will use it in the following way.

_4_ This is different than the most common usage of the term, in which *model* really means "looks like."

___ A model is a device that acts like the thing it is a model of, but is not itself that thing.

___ A model car looks like a car, but it does not run like a car.

___ Our model boat does not look much like a boat, but it does run like a boat.

¶2

___ Modeling is the process of making models that behave like the thing we are modeling.

___ Some models are physical.

___ We can touch them, move them, and measure them.

_5_ A mathematical model for the speed of our boat would give bigger or smaller numbers depending on whether the boat's speed is getting bigger or smaller.

___ The speedometer on your car is, in this sense, a model for your car's speed.

___ The clock in your computer is a mathematical model for time.

___ On the other hand, the spring-driven wristwatch your grandfather used to have is a physical model of time.

___ Other models are mathematical.

APPENDIX B
TEXT RECONSTRUCTION EXERCISES

_2_ Computer modeling is becoming increasingly important to us because in many cases, it is cheaper, quicker, and safer to make a computer model than to make a physical model or to test the real object. ¶3

___ In the 1950s and 1960s, many pilots were killed testing new kinds of airplanes.

___ Today, airplanes are tested as mathematical models before they are even built, and far fewer pilots are killed testing them.

___ Building mathematical models using computers is called *computer modeling*.

Text Reconstruction Key

The paragraphs below show the correct order of sentences for each text reconstruction exercise. When you hand out the initial homework assignment, ask students to number the sentences in each paragraph so as to put them in the correct order. It is also highly beneficial to ask students to rewrite the paragraphs once they have determined the correct order.

Forces, Speed, and Acceleration

Paragraph 1

1. The relationship of force, speed, and acceleration eluded scientists for most of recorded history.
2. Despite impressive mathematical accomplishments, the Greeks never understood the connections.
3. It was not until Newton's efforts less than 350 years ago that a perspective consistent with our modern view emerged.
4. Even then, it took Newton over 20 years to develop his three simple laws.

Paragraph 2

1. How can something as simple as Newton's laws be so confusing?
2. The confusion stems from our common experience of pushing and pulling.
3. In that experience, it takes force to make things move, and more force to make them move fast than to make them move slowly.
4. We therefore conclude that speed and force are related.

Part I

APPENDIX B
TEXT RECONSTRUCTION EXERCISES

Paragraph 3

1. But in Newton's view, speed and force are not related!
2. In Newton's famous second law, force is related to acceleration: the *change* in speed.
3. But why isn't the change in speed related to speed?
4. Here is a way to think about it.
5. How much you eat for dinner tonight is related to how much your weight will increase or decrease.
6. But even if I was told what you ate for dinner every night from the day you were born, I could not figure out how heavy you are.

Paragraph 4

1. Thus in Newton's picture of the world, force tells us about acceleration—the change in speed—but it cannot tell us about speed itself.
2. Yet if I get on my bicycle, I know this is wrong.
3. I have to pedal harder to go faster.
4. But the reason I have to pedal harder is because of air resistance.
5. Instead of pedaling harder, I could reduce my air resistance by getting racing clothes, or by cycling behind an air shield.
6. Then I would go fast without pedaling harder.
7. Thus a cyclist can go faster without increasing the force exerted.

B

Systems

Paragraph 1

1. Systems analysis and systems thinking are increasingly important aspects of the modern world.
2. In any complex situation, there are many related parts that affect one another.
3. Understanding those effects and relations is called *systems analysis*.

Paragraph 2

1. The simplest kind of systems analysis involves understanding whether an increase in one quantity will cause an increase, decrease, or no change in another quantity.
2. If you increase what you eat for dinner, you will increase your weight (unless you also exercise more, or do something equivalent).

Paragraph 3

1. Often the links in a systems analysis can be complicated and even contradictory.
2. If the weather gets colder, you play less baseball and get less exercise.
3. But when the weather gets colder it also snows more and so you go skiing and get more exercise.

Paragraph 4

1. In a simple case, we can tell what will happen without having to calculate with numbers.
2. It snows and I will go skiing.
3. But will I get more exercise than I got back in July?

Part I

APPENDIX B
TEXT RECONSTRUCTION EXERCISES

4. Here I will need to know how many times I played baseball, how many times I went skiing, and how much exercise I got each time.
5. To calculate the exact amount of exercise for each case, I will need to use a mathematical model.

Modeling

Paragraph 1

1. The word *mode*l has several different meanings.
2. We will use it in the following way.
3. A model is a device that acts like the thing it is a model of, but is not itself that thing.
4. This is different than the most common usage of the term, in which *model* really means "looks like."
5. A model car looks like a car, but it does not run like a car.
6. Our case is just the opposite.
7. Our model boat does not look much like a boat, but it does run like a boat.

Paragraph 2

1. Modeling is the process of making models that behave like the thing we are modeling.
2. Some models are physical.
3. We can touch them, move them, and measure them.
4. Other models are mathematical.
5. A mathematical model for the speed of our boat would give bigger or smaller numbers depending on whether the boat's speed is getting bigger or smaller.

6. The speedometer on your car is, in this sense, a model for your car's speed.
7. The clock in your computer is a mathematical model for time.
8. On the other hand, the spring-driven wristwatch your grandfather used to have is a physical model of time.

Paragraph 3

1. Building mathematical models using computers is called *computer modeling*.
2. Computer modeling is becoming increasingly important to us because in many cases, it is cheaper, quicker, and safer to make a computer model than to make a physical model or to test the real object.
3. In the 1950s and 1960s, many pilots were killed testing new kinds of airplanes.
4. Today, airplanes are tested as mathematical models before they are even built, and far fewer pilots are killed testing them.

Part I

APPENDIX C
Sample Answers

Organizing the Data

Activity 3, p. 32

Here is an example of recording and calculating data for the change of speed over time.

Time on Stopwatch (seconds)	Distance From Starting Point (meters)	Speed at Distance (m/s)	Speed Change = $Speed_n - Speed_{n-1}$ (m/s)
0 (start)	0.00	0	0
3	.06		
6	.22		
9	.47		
12	.75		
15	1.07		
18	1.39		
21	1.72		

Students should be encouraged to find their own way to calculate speed, but their thinking needs to be pushed and challenged. Their first ideas are likely to be wrong. Mindless use of speed versus distance/time will generate incorrect answers!

Here are some approaches that work:

1. Graph the data for distance versus time and calculate speed from the slope of the graph. Graphing calculators work well here.
2. Calculate the average speed to that point in time. If the accelerations were constant, speed at the end time would be twice the average speed over the interval.
3. To improve on the second method, calculate the average speed over the time interval. Compare that with the speed at the end of the previous interval and use twice the difference for the speed at the end of the interval. This method only assumes constant acceleration over the time interval, and is quite accurate for small time intervals.

Activity 3, p. 40

Scale Model Extensions Homework

1. Use the weight scale factor to calculate the weight for the model boat based on the Design Brief data for a real boat with no cars or people.
   Scale factor = 12,167,000
   M/V Nantucket = 1,150 tons = 2,300,000 lb
   Model = 2,300,000/12,167,000 = 0.189 lb = 86 g
   Model boat weight = 80 g

2. Estimate the maximum weight of people and cars that the ferry M/V *Nantucket* carries.
   1,100 people × 125 lb/person = 137,500 lb
   60 cars × 3,000 lb/car = 180,000 lb
   People and cars weight = 317,500 lb

3. Determine the scale weight for people and cars in grams.
   317,500 lb/12,167,000 = 0.026 lb = 12 g
   People and cars scale weight = 12 g

Activity 3, p. 43

Minimizing Surface Area

Procedure

3. Compute the total area of the inside surface that touched the sand. Note the units you use for area.
   Method 1: Lay a grid over the "wet area" and count squares.
   Method 2: Divide "wet area" into geometric shapes and calculate area of each shape from length measurements and geometry.

4. What is your conclusion about the relationship between volume and surface area?
   Some shapes have more volume per unit surface area; curved surfaces seem to have the highest.

APPENDIX C
SAMPLE ANSWERS

5. How will this information help you in the design of the boat hull?
   Can help design for reduced friction

Builder Problem

Activity 4, p. 52

Propose *two* distinct methods to minimize fluid friction and describe them in the space below.
One involves reducing wet area; another involves reducing surface roughness.

Design and conduct a test to determine improvements in the friction coefficient for the model boat. If you can, try using the computer model to understand better how the friction coefficient affects the model boat performance. Describe your test below.
There are many solutions. One involves pulling the boat with a spring scale. But be careful: Results depend on speed. It is difficult to pull a boat at a constant speed.

Prototype Construction

Activity 4, p. 53

Proposal

Review student work to look for designs with reduced wet area and designs with smoother finish. Check that their plans are consistent with available fabrication skills and facilities.

Materials

Review student lists for completeness and consistency with any material limitations you have stated.

C

Activity 4, p. 55

Evaluating Your Design

Comparing Performance Data

1. top speed of quick-build = *0.5 m/s*

   top speed of redesigned model = *0.6 m/s*

2. acceleration time for quick-build = *10 seconds*

   acceleration time for redesigned model = *10 seconds*

3. Compare the performance of your redesigned model with that of your quick-build. Provide a brief qualitative description below.
   The new model is slightly faster, about 20%.

4. How would you describe the improvement in performance quantitatively? What calculations do you need to make?
   We will need to repeat baseline measurements and organize the data. We will need to devise a speed versus time graph and compare it with the baseline original.

5. Summarize your results and explain how you have achieved them.
   We reduced water/friction, increasing both the average acceleration and the top speed. We did this by changing the hull shape and by reducing the surface roughness.

Glossary

baseline data: data taken to determine conditions before an experiment is started

beam: the widest point of the hull at the waterline

brainstorming: a group problem-solving technique that involves the spontaneous contribution of ideas from all members of the group

calibration: determining a measurement scale or aligning a device with a measurement scale

centerline: the line of a boat running fore and aft (see keel)

draft: the distance from the waterline to the lowest part of the hull; the draft determines the shallowest depth in which the boat will still float

dynamics: a branch of physics related to the effects of forces

energy conversion: energy in different forms is often converted from one form to another; heat is converted into light, light into heat, motion into heat, and so on.

energy transfer: the movement of energy from one place to another or from one form to another

equilibrium: a balance between two opposing processes such that the net effect of the two processes is no total change. For example, we breathe in about as much air as we breathe out, so that over time, our lungs stay roughly the same size.

factors: the different elements out of which a whole object is made, often used to describe the different causes that lead to a particular outcome

forces: pushes and pulls

friction coefficient: a constant varying with the condition of surface that determines the size of the force of friction. The coefficient is usually measured experimentally because it is too complicated to determine from theory.

frictional force: resistance caused by the passage of water across the hull surface

hull: the outside wall of a ship that begins at deck level and goes down to the keel

GLOSSARY

keel: the centerline of a boat running fore and aft; the backbone of a vessel

performance test: a test to determine how well a design accomplishes an intended purpose

prototype: an original model; the first full-scale and (usually) functional form of a new type or design of a construction

residual resistance: all resistance affecting a body's motion through the water except the frictional force; includes air friction and wave-making

scale factor: a number that indicates how much smaller or larger one object is as compared with a specific copy of it

scale model: a copy of an object (usually smaller) made so that all components are in proportion to one another

scaling: increasing or decreasing the size of something in such a way that all parts are in the same proportion as the original object

section: the shape of a plane passing through the hull perpendicular to the centerline

section coefficient: the ratio of the area of a section divided by the area of a rectangular section having the same beam and draft. It is a measure of the relative fullness of a section and permits the comparison of hulls of differing sizes and shapes.

specifications: the precise details of an invention, plan, or proposal

system: a collection of interacting components that work together as a unit

tolerance: the degree of accuracy demanded in a construction

total displacement: the total weight of a boat including passengers and cargo

variable: an object or quality of changeable value

wetted surface: that part of the hull in contact with the water when the hull is loaded to its total displacement

Part I

Suggested Readings

Bray, A., and C. Curtis. 1994. *The directory of wooden boat builders and designers: A guide to building and repair shops and designers in North America*. Brooklin, ME: WoodenBoat Books.

Chapelle, H. I., and J. Wilson. 1994. *Yacht designing and planning*. New York: W. W. Norton.

Grosslight, L., C. Unger, and E. Jay. 1991. Understanding models and their use in science: Conceptions of middle and high school students and experts. *Journal of Research in Science Teaching* 28 (9): 799–822.

Roberts, N., D. Andersen, R. Deal, M. Garet, and W. Shaffer. 1983. *Introduction to computer simulation: A system dynamics modeling approach*. Reading, MA: Addison-Wesley.

Schauble, L., L. E. Klopfer, and K. Raghavan. 1991. Students' transition from an engineering model to a science model of experimentation. *Journal of Research in Science Teaching* 28 (9): 859–882.

Schuring, D. J. 1977. *Scale models in engineering*. New York: Pergamon Press.

Snir, J., and C. Smith. 1995. Constructing understanding in the science classroom: Integrating laboratory experiments, student and computer models, and class discussion in learning scientific concepts. In *Software goes to school*, ed. D. Perkins, J. Schwartz, M. West, and S. Wiske, 233–254. Oxford, England: Oxford University Press.

Stigler, J., and J. Hiebert. 1999. *The teaching gap*. New York: Free Press.

Winters, J. 1997. *The shape of the canoe*. Bulk Falls, Ontario. *www.greenval.com/jwinters.html*

Part II

CONSTRUCT a CATAPULT

Part II

Introduction

Integrating Science and Technology

Construct-a-Catapult is aligned with the National Science Education Standards (NSES) in physical science, the International Technology Education Association (ITEA) standards, and the National Council of Teachers of Mathematics (NCTM) standards in statistics (see the Standards and Benchmarks Connections table later in this section). Physical science concepts covered in this unit include inquiry and design, dynamics, kinematics, and energy transfer. Through a variety of hands-on design activities, students engage in the iterative processes of scientific inquiry and technological design.

Schedule and Cost

The time needed to complete the core activities is 13 class sessions, or about two-and-a-half weeks, with material costs totaling about $90 per class of 25 students:

- Design Brief and Quick-Build = $30 or less for a class of 25 working in groups of three or more
- Research = $10 or less
- Development = $30 or less for a class of 25 working in teams of five
- Communication

Key Ideas

Three of the key science ideas (elasticity, energy, and force) used in Construct-a-Catapult are covered in text reconstruction exercises in Appendix B (p. 179). The first exercise, on force, involves a very simple reconstruction and is intended as an introduction to text reconstruction. The other exercises are more difficult.

Elasticity

Hooke's law states that the elastic property of matter is quantifiable for specific materials. Students measure force/stretch relationships of common rubber bands as the basis for a reasonably accurate projectile delivery system.

Energy

The conversion of elastic potential energy into kinetic energy is used to determine range.

Force

Force is studied and measured in the context of Hooke's law and principles of gravity.

Calibration

Students calibrate and document procedures for their prototype to meet performance specifications.

Inquiry and Design

The design process involves students in inquiry, generating solutions, development, testing, and evaluation. Readings on the history of ancient catapults (c. 400 BC) give students a context for the background of a catapult's technological design.

Student Portfolios

The following items can be accumulated in portfolios for summative assessment.

INTRODUCTION

- Pretest: Snapshot of Understanding
- Initial questions: Design Brief
- Individual information search: Catapult Design History
- Quick-build sketch
- Brainstorming record: Identifying Variables
- List of variables: Beyond the Quick-Build
- Elasticity investigation design and results: Investigating Elasticity
- Group process description: Design Process
- Part name labels: Parts of a Catapult
- Fastener selections: Resource List: Fasteners and Adhesives
- Beyond the Quick-Build: Second Pass
- External feedback record: Team Feedback Form
- Landing pattern analysis: Making a Frequency Distribution
- Force-distance graph: Making a Launching Graph
- Postchallenge reflection: Reflections on Your Design
- Prototype demonstration notes: The Challenge Event
- Group summary documentation: Creating a User's Manual
- Posttest and self-assessment: Snapshot of Understanding
- Oral presentation (optional)

INTRODUCTION

Standards and Benchmarks Connections

Task	Source
Students conduct independent information searches in the context of human inventiveness. **Standard/Benchmark:** Issues in Technology; Technology and Society; Science, Technology, and Society	AAAS 9–12, ITEA V, NCSS VIII
Students build to specifications, manipulate, and observe interactions of parts in operation of a simple elastic mechanism. **Standard/Benchmark:** Systems; Technology and Society; Evidence, Models, and Explanation	AAAS 9–12, ITEA 9–12, NSES K–12
Students describe variation and identify variables and corresponding potential controls for improving design to meet performance specifications. **Standard/Benchmark:** Systems; Nature of Technology; Evidence, Models, and Explanation; Problem Solving	AAAS 9–12, ITEA 9–12, NCTM 1, NSES K–12
Students use elements of inquiry and investigate elasticity to inform design. **Standard/Benchmark:** Design and Systems; Nature of Technology; Science as Inquiry	AAAS 9–12, ITEA 9–12, NSES 9–12
Students use Newton's second law of motion together with Hooke's law to quantify component performance parameters. **Standard/Benchmark:** Motion; Nature of Technology; Technology and Society; Physical Science; Constancy, Change, and Measurement	AAAS 9–12, ITEA 9–12, NSES 9–12 and K–12
Students create tables and represent data in appropriate graphic format. **Standard/Benchmark:** Critical-Response Skills	AAAS 9–12
Students communicate, orally and in writing, their interpretation of this investigation and what variables to control in design development. **Standard/Benchmark:** Scientific Inquiry; Nature of Technology	AAAS 9–12, ITEA 9–12
Students apply abilities of iterative technological design, including brainstorming, research, ideation, choosing among alternative solutions, development, implementation, and evaluating consequences. **Standard/Benchmark:** Systems; Nature of Technology; Technology and Society; Science and Technology; Systems, Order, and Organization	AAAS 9–12, ITEA 9–12, NSES 9–12 and K–12
Students utilize tools and processes to construct and modify working models. **Standard/Benchmark:** Nature of Technology	ITEA 9–12
Students collect, represent, and statistically process test data to calibrate their design prototype. **Standard/Benchmark:** Nature of Technology; Statistics	ITEA 9–12, NCTM 10

INTRODUCTION

Task	Source
Students create procedural operating instructions for others to use. **Standard/Benchmark:** Mathematics as Communication	NCTM 2
Students interpret and follow directions in the challenge demonstration event. **Standard/Benchmark:** Nature of Technology	ITEA 9–12
Students communicate quantitatively the technical performance specifications and operating instructions for their prototype. **Standard/Benchmark:** Communication Skills; Nature of Technology	AAAS 9–12, ITEA 9–12
Students articulate principles of science employed in a catapult's operation. **Standard/Benchmark:** Critical-Response Skills; Science and Technology	AAAS 9–12, NSES 9–12
Students self-assess their learning by comparing pre- and post-Snapshots of Understanding. **Standard/Benchmark:** Issues in Technology; Science, Technology, and Society	AAAS 9–12, NCSS VIII

Source Key:

AAAS = American Association for the Advancement of Science. 1993. *Project 2061: Benchmarks for science literacy*. New York: Oxford University Press.

ITEA = International Technology Education Association. 1996. *Technology for all Americans: A rationale and structure for the study of technology*. Reston, VA: ITEA.

NCSS = Task Force on Social Studies Teacher Education Standards. 1997. *National standards for social studies teachers.* Washington, DC: National Council for the Social Studies.

NCTM = National Council for Teachers of Mathematics. 1991. *Professional standards for teaching mathematics.* Reston, VA: NCTM.

NSES = National Research Council. 1996. *National science education standards.* Washington DC: National Academies Press.

Part II

INTRODUCTION

Course Outline

Introduction

- Overview: Catapult Design Brief
- Snapshot of Understanding
- Catapult Design History

Quick-Build

- Overview: Quick-Build Catapult
- Identifying Variables
- Beyond the Quick-Build

Research

- Overview: Research
- Investigating Elasticity

Development

- Overview: Development
- Parts of a Catapult
- Resource List: Fasteners and Adhesives
- Beyond the Quick-Build: Second Pass
- Team Feedback
- Making a Frequency Distribution
- Making a Launching Graph
- Reflections on Your Design

Communication

- Overview: Communication
- Creating a User's Manual
- The Challenge Event: Usability Testing
- Snapshot of Understanding

DAY 1

ACTIVITY 1
Catapult Design Brief

OVERVIEW: CATAPULT DESIGN BRIEF

In this unit, you will be designing, building, and improving a mechanical launching system resembling an ancient catapult. Your system will be scaled down for use in the classroom and will take advantage of elastic properties of modern materials, saving you tasks like twisting huge bundles of sinew into torsion springs (see Catapult Design History reading). You will use both technological design and scientific inquiry as processes to investigate and improve how your catapult performs.

Design Challenge

As a member of a product development team, you are to design, build, and document a mechanical launching system that can deliver a small object predictably and repeatedly over a specified range of distances.

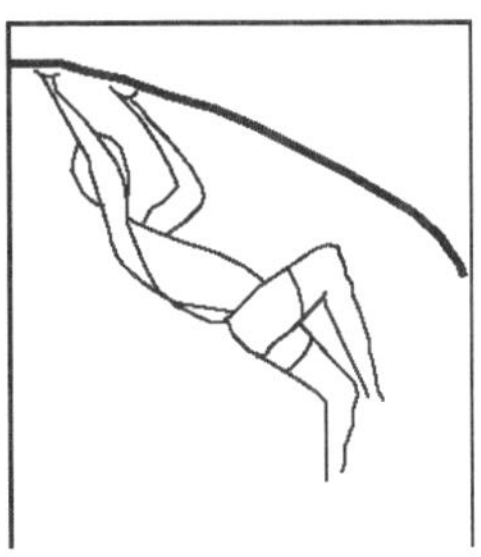

The material property of elasticity used in the catapult is used today by pole-vaulters. Since the inception of the flexible fiberglass pole, world records have increased by more than 2 meters.

Scope of Work

- *Quick-Build:* Build a design according to specifications.
- *Research:* Investigate elasticity, and identify variables you can control to create a more accurate catapult.
- *Development:* Redesign, build, and test; collect data and analyze patterns of results; then calibrate your prototype catapult with a launching guide and graph to meet the challenge.
- *Communication:* Produce a user's manual that documents your design and its operation, including sketches, charts, launching graphs, and notes.

Key Questions

Write two questions that you have regarding your challenge and/or scope of work.

Part II

ACTIVITY 1
CATAPULT DESIGN BRIEF

> **Safety Alert!**
> In this exercise you will be launching elastic bands. You must wear safety glasses or goggles whenever elastics are launched and be careful to aim away from others.

SNAPSHOT OF UNDERSTANDING
What I Already Know About Catapults

The unit of study you are about to begin will challenge you to design, build, test, and calibrate a working model of a mechanical system most closely resembling an ancient catapult. Controlling the system will require you to investigate elasticity. Before you begin, record what you already know about catapults, forces, and energy by answering the questions below. This is not a test—it is a series of questions that ask about your current knowledge of the material in this unit. At the end of the unit you will answer similar questions, after which you can compare what you have learned with the answers you give here.

1. Using a rubber band as a catapult and a pie plate as a target, try to shoot a rubber band into the pie plate placed on the floor. Move the pie plate closer to you or farther away and keep trying to shoot the rubber band into it.

    a. What important variables do you need to deal with as you try to hit the pie plate?

    b. What scientific principle(s) can you use to describe how your elastic catapult operates?

1

SCiLINKS®
THE WORLD'S A CLICK AWAY

Topic: force
Go to: *www.scilinks.org*
Code: CAC01

Topic: energy
Go to: *www.scilinks.org*
Code: CAC02

2. What does a catapult look like? Make a sketch and label its parts or describe in words in the space below.

3. List any current uses of catapult-like devices of which you are aware. (For example, catapults are used to launch planes from aircraft carriers.)

4. Have you ever designed and built anything? (*yes/no*) If yes, describe how you went about your designing and building. If not, think about how you might go about designing and building something and describe below.

5. From your previous learning or experience, describe what the following terms mean.

force:

energy:

Part II

ACTIVITY 1
CATAPULT DESIGN BRIEF

Catapult Design History

Throughout history, humans have applied innovative ideas and designs to devices for throwing weapons. First, the sling was developed to overcome the limitations of the human arm. Next, hunters and soldiers devised the bow and arrow to improve aim and velocity. Eventually, major advances in power and accuracy were achieved with the design of machines called *catapults*. Early catapults were modeled after the bow and arrow, but they quickly evolved into strong, single-armed machines constructed of composite layers of wood, sinew, and horn. This new weapon for attackers unbalanced the advantage once held by defenders during enemy siege (see The Catapult Advantage section on p. 114). While the defenders still had the ability to prepare for attack by building large walls, attackers—using catapults—could physically overcome these obstacles. These accurate machines also provided cover fire for troops attempting to breach enemy walls.

> *Catapult* is derived from the Greek prefix kata, denoting downward motion, and from pelte, a light shield carried by Greek troops. A katapelte could smash a projectile downward completely through a shield.

From Tension Bow to Torsion Spring

The first catapults were designed under the direction of Dionysius the Elder, ruler of the Greek colony of Syracuse (in what is now Sicily) in 399 BC. To prepare his city for a long war with Carthage, Dionysius assembled large research and development teams to create products that would give Syracuse strategic advantage in the upcoming war. The teams were made up of specialists who divided their labor into manageable units. Research and development enterprises still use this practice today.

Under Dionysius's direction, Greek artisans created the *gastrophetes*, or "belly bow," modeled after the bow and arrow. To cock the weapon, the archer pulled the stock—the compound beam forming the main axis of the weapon—into his abdomen and pulled back the string with both arms. Using two arms to cock the bow created substantially more power than the traditional handbow, for which the archer used only one hand. This tension bow was also larger than the handbow and consequently was able to hurl heavier arrows. However, the gastrophetes lacked the ability to throw arrows more than 300 yards and was incapable of throwing stones.

To address the shortcomings of the gastrophetes, further research led to the development of a new type of bow: the *ballista*. This bowlike device was made of

1

Torsion is the reactive torque (rotational force) that an elastic exerts as a result of being twisted on its axis. Tightly twisting and then releasing two strands of rope yields rapid untwisting, demonstrating torsion.

two independent lateral arms connected by a bowstring at the outer ends. Unlike earlier bows, the ballista used the power of torsion to propel stones. Bundles of cord or animal sinew were twisted to energize the bowstring; more twisting created a greater torsion effect and therefore more power. Although similar in form to the gastrophetes, the substitution of torsion for tension made the ballista a more powerful machine.

Ballista

By the fourth century AD, the latest projectile-launching device was the one-armed, torsion-driven, sling machine known as the *onager,* or "wild donkey," named for the bucking action it exhibited under the force of recoil. Unlike the bows used in earlier times, the onager combined both throwing and slinging motions, which extended the weapon's range. A single arm extended from the torsion bundle and ended in either a cup or a sling, which held the stone. The addition of a sling to the arm of this catapult increased its power by at least a third and allowed the machine to hurl a missile in a high arc—over potential obstacles—toward a target. Today the onager is the weapon most people associate with the term *catapult*.

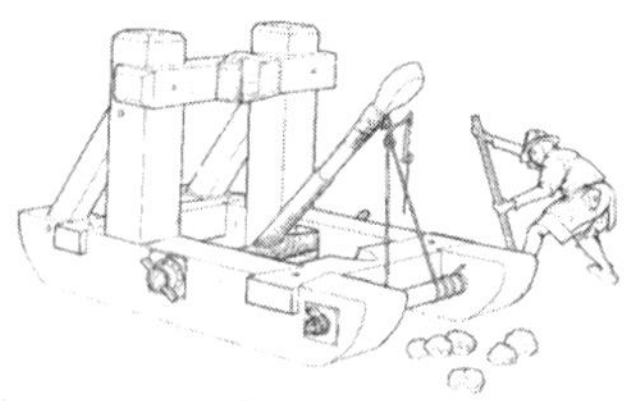

Onager

Falling Weight Devices

By the end of the sixth century, a new stone projector called the *traction trebuchet* had appeared in the Mediterranean. The traction trebuchet was a medieval catapult-like device that threw missiles with the force of up to 250 men. Nearly all catapults used at this time operated by a sudden release of energy; an exception was the medieval *counterweight trebuchet*. Similar in action to a seesaw or a slingshot, this trebuchet

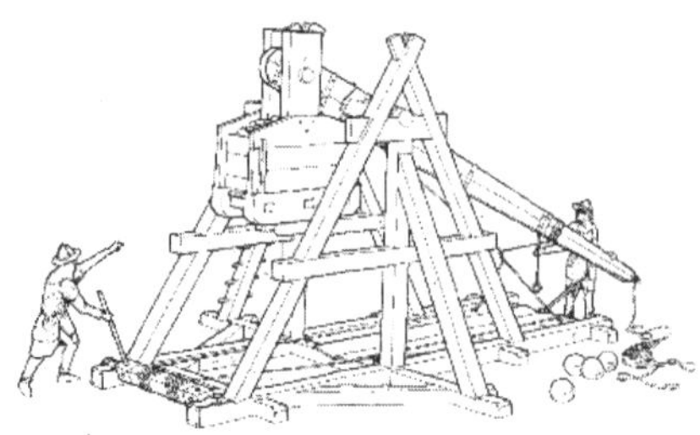

Trebuchet

Part II

ACTIVITY 1
CATAPULT DESIGN BRIEF

used the energy of a falling counterweight that was suspended from one end of a wooden arm. This propelled a missile that was placed in a sling at the other end of the arm. These machines were simpler to construct, operate, and maintain than those with sinew torsion bundles. Trebuchets were used throughout the Middle Ages and up through the siege of Gibraltar by the French and Spanish fleets in 1779–1782.

SCILINKS THE WORLD'S A CLICK AWAY

Topic: elasticity
Go to: *www.scilinks.org*
Code: CAC03

Topic: mechanical advantage
Go to: *www.scilinks.org*
Code: CAC04

Evolution of Catapult Design

Beginning with the ballista, early catapult engineers combined several design elements to simulate an archer's motions for consistent and accurate targeting. These included

- *cams* to transfer rotating motion into sliding motion;
- *claw-and-triggers* to grasp and release the bowstring;
- *dovetail grooves* with *sliders* to form a movable joint;
- *flat-link chains* to connect other design elements;
- *pedestals* to provide stability and support;
- *ratchets and pawls* or *winches* to allow incremental accumulation of applied human energy for increased power;
- *stocks* to form the main axis of the weapon;
- *torsion springs* to store the energy used to propel an object; and
- *universal joints* to allow rotation.

A number of design features that made their debut on catapults are still in use today. These features, and examples of modern uses, include

- *cams* in racecars;
- *sliding dovetail surfaces* in woodworking;
- *flat-link chains* in necklaces and conveyor belts;
- *torsion springs* in garage doors; and
- *universal joints* in automobiles and aircraft.

In addition to making significant advancements in military technology, catapult engineers used experimental procedures, derived optimization and scaling formulas, and performed advanced calculations that showed a level of engineering rationality not achieved again until the time of the Industrial Revolution of the 18th and 19th centuries.

The Catapult Advantage

Before the development of catapults, the strategic advantage in ancient warfare was held by a defending army, who fought behind walled cities. An attacking army armed with a catapult, however, was no longer at a disadvantage. The ability of the catapult to concentrate hits on a single spot rendered defensive wall battlements and armored siege towers vulnerable, changing forever the equilibrium of politics and society.

Even after the invention of cannons and mortars in the late Middle Ages, catapults were still active on the battlefield because they were easy to construct on site and were able to do a great deal of damage with more reliable results than the inconsistent gunpowder of the day. They had the further advantage of positioning flexibility and relative noiselessness.

Today catapults can be seen in medieval reenactments, engineering contests, and period films such as *Robin Hood.* Movies and television have even used catapults to produce dramatic stunts—such as flying livestock in *Monty Python and the Holy Grail* and an airborne piano in an episode of *Northern Exposure.*

Part II

DAY 1

ACTIVITY 1 Teacher Pages

OVERVIEW: CATAPULT DESIGN BRIEF

Give students the Catapult Design Brief and the Snapshot of Understanding. Begin a class discussion highlighting the important design issues they should consider. For homework, students read about the history of catapults.

Design Challenge

Have students read the design challenge, and explain that they will use processes in technological design and scientific inquiry together to meet the challenge. At the end of the unit, another team will operate their prototype in a final Challenge Event, using a user's manual written by the prototype developers. For this reason, it is essential that the team write a user's manual that effectively communicates the important elements of their team's experience with the catapult.

Snapshot of Understanding

Students begin by shooting several rubber bands into a pie plate, using an "insta-pult" developed in a fast-paced introductory activity. They then write short answers to questions about their prior knowledge of catapults, the design process, and underlying science concepts. The insta-pult adds a performance dimension to this assessment that will increase with student interest.

Catapult Design History

Students read for homework an illustrated summary of catapult design in historical context.

Preparation

- Read and become familiar with the entire unit.
- Photocopy student activity sheets for distribution.
- Obtain insta-pult materials: rubber bands and tin pie plates.
- Devise rules of safety and conduct for the insta-pult.

- Define your assessment system with a clear, simple description.
- Preview library and internet resource information.

Materials

- Student Activity Sheets
    - » Course Outline
    - » Overview: Catapult Design Brief
    - » Snapshot of Understanding
    - » Catapult Design History
- Ring, pocket, or folio binder (student supplied; for keeping student activity sheets, notes, and drawings for reference and user's manual)
- safety glasses or goggles
- rubber bands
- tin pie plates

Time Requirement

This activity requires one class session. It takes about 20 minutes to complete the Snapshot of Understanding.

Teaching Suggestions

Hand out the Catapult Design Brief student activity sheets. Ask students to keep these and future sheets together and to bring them to the classroom with other notes to serve as a record and reference for daily activity (and assessment) in the unit. Advise students that they will work in teams, using processes of technological design and scientific inquiry, and that other teams will critique their prototype with respect to the challenge criteria.

Students are also required to document their activity in a laboratory journal in order to contribute effectively to the final team presentation and to enhance their

ACTIVITY 1 TEACHER PAGES

individual portfolios. Be clear on your rubrics for assessing their work and share them with students. Indicate which activities will be individually graded and which will be given a team score. Be prepared to justify team scoring if some students (or parents) are not used to the idea.

Preassessment

Hand out the Snapshot of Understanding. Emphasize that it is not a test and that students will not be graded on this activity. The purpose of the Snapshot is self-diagnostic: to find out what students know initially about the key science and technology learning objectives of Construct-a-Catapult.

An inventory of students' prior knowledge is an important teaching and learning tool. Not only does the inventory help guide students toward the concepts they need to learn the most, but it also prepares them to accept new information in a manner that ties meaningfully to what they already know. At the end of Construct-a-Catapult, students will be able to compare these answers with answers to similar questions given at the end of the unit.

Collect and retain the Snapshot after students complete it.

Issuing the Catapult Challenge

Refer the class to the Catapult Design Brief and discuss the challenge statement. Although it is important to spend some time discussing the challenge, it is important to move ahead rapidly and begin the introductory insta-pult activity.

Contextual Design History Reading and Internet Search

Assign the *Catapult Design History* reading for homework. You may wish to extend student investigation of the history of catapults to include internet searches.

DAYS 2

ACTIVITY 2
Quick-Build Catapult

OVERVIEW: QUICK-BUILD CATAPULT

Your first step is to build a simple model catapult for launching a projectile. You will be limited in time and materials to do this. The general design may be determined by your teacher or by your creativity with the available materials. The idea is to build this experimental launcher as quickly as possible and try it out to get a sense for what is important in building your next catapult, which will be designed for accuracy.

Once you have put it together, take your quick-build to a designated testing area and try launching the projectile. Your goal is to find ways to predict and control where the projectile goes. Each team partner should do some launching and take a turn at observing, making notes, and retrieving projectiles. Here are some suggestions:

- Put the tin pie plate target somewhere in the launching area and try hitting it repeatedly.
- Move the target to find a minimum and maximum distance (*range*) over which you can achieve some measure of control.
- Use books or boxes to elevate one end of the catapult.
- Experiment with different ways of hitting the target, such as low and direct versus a high lob.

As you are building and testing, jot down notes so that you can refer to them later. Use your trial launch observations with the quick-build catapult to identify variables that must be controlled to meet the challenge of predictable and repeatable (accurate and reliable) performance.

- What did you see happening?
- What did you change?
- What happened as a function of the change(s)?

ACTIVITY 2
QUICK-BUILD CATAPULT

Identifying Variables

Your catapult is a system of interrelated parts. You may have observed that making a change in the catapult itself or in the way you operate it can make a difference in the catapult's performance. Think of the variables you observed as belonging to one of two categories: (1) parts of the catapult system or (2) user operating procedures. In the table below, list these variables by category and describe the range of possible variation. Focus on detail and try to think of factors you can control or modify to improve the accuracy of your launcher. Use additional paper as necessary.

Parts of the Catapult System	
Variable	**Range of Variation**
Example: *Angle of the catapult*	*Flat, between 0° and 45°, between 45° and 90°*
User Operating Procedures	
Variable	**Range of Variation**
Example: *How you position the rubber band*	*High on the nails, in the middle of the nails, low on the nails*

2

Beyond the Quick-Build

From your list in the previous activity, select the variables that you feel contributed most (positively or negatively) to the performance of your quick-build. Suggest what modifications you could make to meet the challenge goals. Record your current thinking below:

Most Significant Variables	Suggested Modifications

Part II

2 DAYS

ACTIVITY 2
Teacher Pages

OVERVIEW: QUICK-BUILD CATAPULT

In a single, fast-paced class session, students work in teams to design and construct a materials-constrained quick-build catapult. Each team must be composed of at least three students to complete the testing phase successfully. Students use their quick-build to practice firing uniform projectiles (such as practice golf balls or clay balls) at a target to determine factors important to accuracy. They record notes on the changes they make and on the performance achieved, and they sketch a drawing with labeled parts. During homework and class discussion, they identify important variables.

Identifying Variables

For individual homework, students reflect on their quick-build performance and brainstorm a list of system and operational variables observed, with ranges of variation. They contribute their list to a class compilation and discussion.

Moving Beyond the Quick-Build

After class discussion, student product development teams select those variables deemed most significant to the challenge performance goals and outline design features to control them.

Preparation

- Tailor quick-build instructions and assembly drawing to materials obtained; copy and distribute.
- Select a uniform projectile and provide to all teams. Carefully weighted lumps of clay have the advantage that they will not bounce, but they require extensive preparation and maintenance. Other options include practice golf balls, small beanbags, or pincushions.
- Designate and arrange construction and launching areas for the quick-build.
- Organize materials for easy access.
- Determine student team sizes and formation strategies.

- Contact related workplace career representatives in science, engineering, history, military, avalanche control, and other related areas to establish context and reference.

Materials

For Each Student:

- Student Activity Sheets
  - » Overview: Quick-Build Catapult
  - » Identifying Variables
  - » Beyond the Quick-Build

For Each Team:

Building Materials

- scrap board or Peg-Board
- 2–4 screws, nails, or bolts, 2–5 cm long
- rubber bands of various lengths
- short pieces of string or twine
- projectiles (clay, practice golf balls, small beanbags, or pincushions)
- tin pie plate target

Tools

- awl, punch, nail, or drill for fastener pilot hole
- screwdriver or hammer
- ruler showing millimeters
- pencil
- safety glasses or goggles
- disposable camera with a flash

ACTIVITY 2 TEACHER PAGES

Time Requirement

This activity usually requires two class sessions: one to build and test the quick-build catapult and one to discuss variables and plan modifications.

Teaching Suggestions

Hand out student activity sheets, and remind the students to keep all sheets together and bring them to the classroom with other notes to serve as a record and reference for daily activity (and assessment) in the unit. These records will be critical to their successful development of the final team's user's manual, as well as their individual portfolio.

Building, Testing, and Sketching the Quick-Build

This needs to be an active, fast-paced class. Allow students only one class session to build and test a quick-build launcher. Divide students to work in groups of three or four—refer to these groups as product development teams. You can leave the design wide open or suggest a very simple design such as that shown in the Parts of a Catapult activity sheet (p. 140). If you want students to assemble by reading a dimensioned drawing, you may wish to dimension or otherwise modify the Sample Student Catapult Designs (p. 157–158) to match your supplies.

Arrange the common tools, such as hammers and screwdrivers, so that students have easy access to them. Students who have rarely used a hammer or screwdriver may need a demonstration of safe and effective use. Intervention by you and by "student consultants" at critical junctures may also be required.

Each student should make a rough sketch of his or her group's quick-build and for homework fill in labels for all parts. If appropriate for your class, you might recommend that manual drafting, computer-aided design (CAD), or another technical illustration process be used for later inclusion in the user's manual.

Homework: Identifying Variables

Set the stage for design improvement analysis with the Identifying Variables activity sheet. Students list all the variables they observe as they build and test their quick-builds, and specify for each the range of possible variation.

Students are asked to classify the variables in two categories: those that are part of the device itself and those more associated with the user. An example of each kind of variable is included on the student activity sheet. Listed below are some variables you might expect students to identify:

Parts of the Catapult System

- kind of rubber band
- tightness of the rubber band
- angle at which the catapult is positioned
- stability of the catapult's base
- how far back the rubber band is pulled
- how far apart the nails are positioned

User Operating Controls

- how the user handles the projectile
- how the user positions the rubber band on the nails
- where the rubber band is held
- consistency of operator stance, steadiness, and concentration
- smoothness of release

ACTIVITY 2 TEACHER PAGES

Class Discussion of Variables

Compile a class list of the variables and ranges of variation students identified in their homework. You may want to make a large chart on the board and jot down students' ideas so that all students have access to the complete list. Use a class discussion to set the stage (and to level the playing field) for the next activity.

Planning Modifications

The product development teams should begin their work with the Beyond the Quick-Build activity sheet. (A team must be composed of at least three students: one to launch, another to mark landings, and a third to record test data.) Teams discuss and select those variables deemed most critical and controllable for success in meeting the challenge. They record, in words or sketches, their plans to modify or redesign a launcher to address the problems.

Part II

3 DAYS

ACTIVITY 3
Research

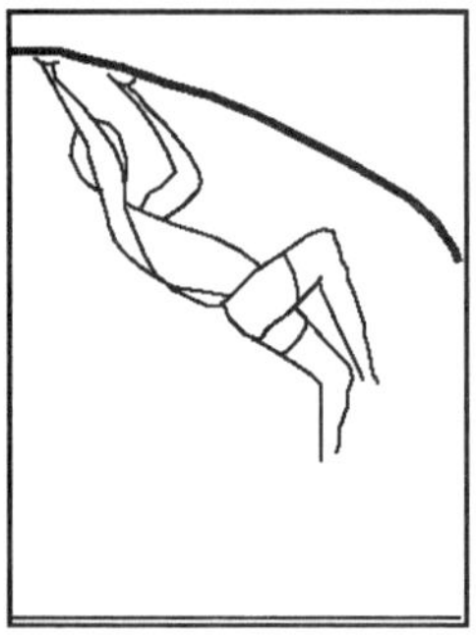

The material property of elasticity employed in the catapult is used today by pole-vaulters. Since the inception of the flexible fiberglass pole, world records have increased by more than two meters.

OVERVIEW: RESEARCH

You have assembled and fired a simple, inaccurate quick-build catapult. You probably observed inconsistency in performance (unpredictable landings) from your rubber band–powered launcher. To make successful improvements on this device, you will need to learn about sources of error (nonrepeatable results) in both the mechanical system and the human operating technique. You will focus first on basic research regarding elasticity, using your actual quick-build in the experimental setup. You will work in pairs or teams to conduct the research activities.

Scope of Work

- Research the science concepts relating Hooke's law to the functioning of the quick-build catapult.
- Suspend known mass loads from your quick-build's rubber band(s).
- Tabulate and graph stretch as a function of applied mass.
- Use Newton's second law of motion and the gravitational constant to derive force/stretch calibration values for the rubber band(s). This will tell you how far to pull the rubber band to achieve a certain range.

As you make design improvements to meet the challenge conditions, you may need to come back to these basic research methods to test and calibrate new configurations of the elastic propulsion system. No matter how carefully you predict and plan, there may be mistakes or surprises; be alert and learn from them so that you can modify your experiments accordingly.

Part II

ACTIVITY 3
RESEARCH

Investigating Elasticity

A key part of your catapult design is the elastic propulsion system. In the quick-build activity you experimented with common, convenient rubber bands. Did you notice any useful relationships between the force you applied, the amount of stretch of the rubber band, and the distance the projectile traveled? Did the projectile go farther if you pulled harder to stretch the band? Did you notice any limits or exceptions?

This activity will help you discover and quantify the relationships between the elasticity of the rubber bands, the force applied to them, and the distance they traveled. Matter has some universal elastic properties, whether it be rubber bands, concrete, or steel (skyscrapers, which are made of steel and concrete, sway back and forth several feet during high winds). These properties were observed and described by Robert Hooke (1635–1703) in his "hypothesis of springiness." A contemporary of Newton, Hooke discovered what he believed to be a law of nature similar to Newton's laws of motion.

Although Hooke's law is an important discovery, it is different in scope and character from Newton's laws. In the design of your catapult, you will look at how Hooke's law and Newton's second law of motion are connected to each other and discuss the differences in their application.

SCILINKS®
THE WORLD'S A CLICK AWAY

Topic: force
Go to: *www.scilinks.org*
Code: CAC01

Topic: elasticity
Go to: *www.scilinks.org*
Code: CAC03

Topic: gravity
Go to: *www.scilinks.org*
Code: CAC05

Topic: mass
Go to: *www.scilinks.org*
Code: CAC06

Experimenting With Stretch

1. Select two rubber bands that are different in width, length, thickness, stiffness, and so on. Name and label them, and record their dimensions and other distinguishing characteristics.

   Band A:

   Band B:

2. For each rubber band, attach both looped ends to projections as shown below so that units of applied mass ("weights") may be attached and hang freely. You may wish to secure your quick-build catapult vertically on the edge of a table or window ledge.

Horizontal view without weights:

Vertical view with weights hanging:

3. a. Select a series of 5–8 known masses to suspend from each rubber band. Your least mass must be sufficient to produce measurable stretch.
   b. Tape a paper strip to your quick-build board and mark the vertical location of the unweighted rubber band as the starting point for your measurements.
   c. Using paper clips or other hooks, attach each mass in sequence from least to greatest to the suspended rubber band, and measure the amount of stretch.

4. For each rubber band, record the applied masses in kilograms and corresponding distance stretched in a table of data points. Plot your data as graphs similar to those shown below. For each graph draw the best-fit line through the points. Tape each rubber band to the legend of its graph.

Band A

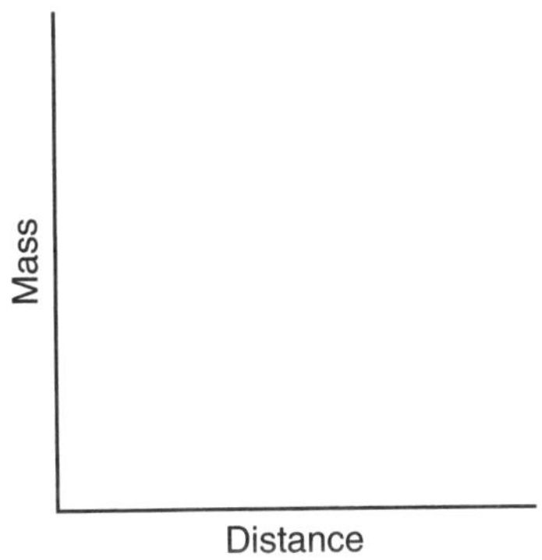

Band B

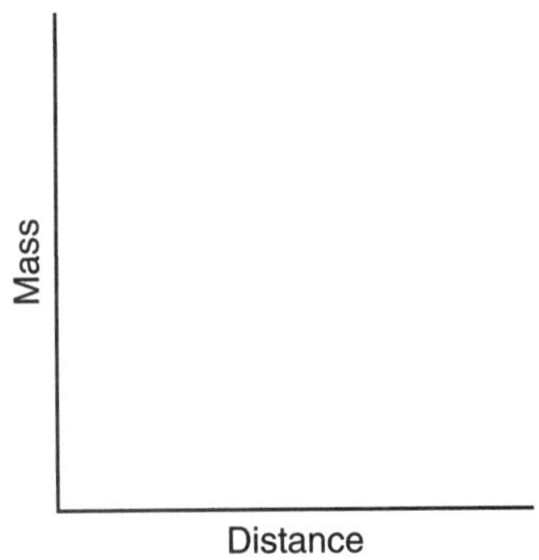

5. a. Describe what you noticed about the points plotted for each rubber band.

   b. Could a straight line include all or most of your data for each rubber band? Explain.

6. Use your graph to predict the distance of stretch for masses smaller and larger than those you used.
   a. How accurate do you think these predictions will be? Explain.

   b. Pick a mass between that of two you used. For example, if you had used 45 g and 60 g in your calibration, choose a mass in between—perhaps 55 g. Figure out the length of the stretch corresponding to this applied mass.

   c. What do you think happens in the case of very large applied masses? Explain.

7. Try to express the relationship between mass and distance as a formula.

8. Hooke's law states that the stretch is proportional to the mass you applied. The region in which a material—in this case the rubber band—obeys Hooke's law is called the *elastic limit* of the material.
   a. What does your graph tell you about the elasticity of your rubber bands?

   b. How are your bands' elastic limits represented on your graph? Can you find a region on the graph where Hooke's law is not followed?

9. a. Plot both lines from the graphs you made in question 4 on the same set of coordinates.

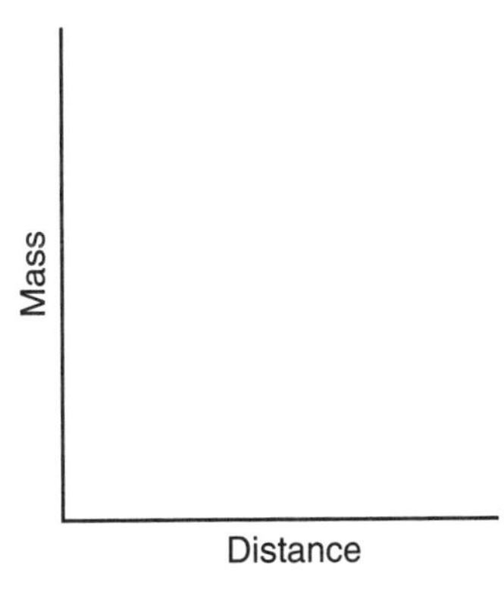

b. Compare the two lines and explain similarities and differences between them. Be sure to compare them with regard to their conformity to expectations from Hooke's law.

c. What do you think the area under your plotted lines represents?

10. How could this exercise help you design your mechanical launching system?

Part II

3 DAYS

ACTIVITY 3 Teacher Pages

OVERVIEW: RESEARCH

Students work in pairs or teams to investigate elasticity. They suspend known mass loads from rubber bands in an experimental design that uses their quick-build launchers. Students then create a data table and a graph to show the amount of stretch as a function of applied mass. They use the gravitational constant and Newton's second law of motion to convert the mass loads applied to the rubber bands on their catapults into force calibration values; they also relate their data to Hooke's law. In calibrating the rubber band, students are creating a measurement scale that relates the amount of stretch to the quantity of force (and hence the range the projectile will travel). As an option, you may consider graphic or mathematical derivations of spring constants and energy factors.

Preparation

- Devise ways of affixing quick-builds on edge vertically for the experimental setups.
- Determine suitable type, quantity, and variety of mass loads for appropriate stretch of rubber bands.

Materials

For Each Student:

- Student Activity Sheets
    - » Overview: Research
    - » Investigating Elasticity
    - » Projectile Motion (optional)

For Each Pair or Team:

- clamps or duct tape for stabilizing each quick-build in vertical position for mass loading
- four to six types of rubber bands, about 25 of each type (ways in which they can differ include their width, length, thickness, and stiffness)
- set of masses
- graph paper
- paper clips or hooks for suspending masses

For the Class:

- Accurate scale or balance for quantifying mass load values in grams

Time Requirement

This unit requires three class sessions:

1. One class for experimental setup, mass loading, and stretch measurement for two different rubber bands. You might assign the graphing of data for homework.
2. One class for whole-group comparison of graphs and merging separate graphs to a single set of axes.
3. One class for discussion of concepts, formulas, and steps for conversion from suspended mass to spring force loading in the context of Hooke's law, applied to a quick-build catapult propulsion system.

ACTIVITY 3 TEACHER PAGES

Teaching Suggestions

Day 1

Ask students to work in pairs on the Investigating Elasticity activity. Attach the quick-build launchers to a vertical surface so they can be used as elastic supports.

Students will suspend masses in increasing increments from rubber bands and create a data table of mass versus stretch for two different rubber bands. Comparing findings across the class should lead to appreciation of the numerous variables inherent in the elastic component of the propulsion subsystem.

Using the quick-build in the experimental setup makes the data directly relevant to subsequent redesign activity. Students should complete data collection and organize their results in a table during this class session. They will then construct a graph for each set of data in the table and answer the series of homework questions in the activity sheet.

Day 2

Review homework by choosing several groups of students to sketch or project their graphs on the board for use as a basis of discussion. Ask them to display the rubber bands associated with each graph. To assess the level of class understanding, pose questions and elicit responses from a range of students. Sample questions might include:

- *What do these graphs tell us about the elastic behavior of the rubber bands?*
- *Why are the graphs different?* (Relate to observed properties of each elastic band, such as length, width, thickness, and stiffness.)
- *How are the bands' elastic limits represented on these graphs?*

A straight line will describe the data within each band's limits of elasticity. The students' own findings should set the stage for your presentation of the formula for Hooke's law and discussion of elastic limits. A change in line slope, or the beginning of curvature, indicates the deformation zone, where stretch is partially irreversible.

Hooke's law states that force is proportional to stretch and is expressed by the formula $F = kd$. This is sometimes written as $F = -kd$ to show that the force is in the opposite direction as the stretch. Students are often confused by the fact that this formula is sometimes written with a minus sign and sometimes not. The minus sign is important when we are concerned with the direction of the force. When the stretch is no longer proportional to force (when twice the force produces less than twice the stretch), you have exceeded the elastic limit of the rubber band. You have also permanently changed its properties.

Correlations between Newton's second law and Hooke's law should be explored in the second day of this activity. On the first day, students should have processed their observations somewhat subjectively, gaining experience in how to select an elastic band for their next launcher's propulsion system that will remain within elastic limits. Students should come to understand that if they load their catapult's rubber band beyond its elastic limit, the permanent deformation will change subsequent performance, thus rendering calibrations useless.

For the remainder of the session on Day 2, guide student groups in choosing an appropriate scale such that their graphs fit onto one set of axes. Students should complete the Investigating Elasticity activity sheet in class. Answers to questions 9 and 10 should reflect the outcomes of the previous in-class discussion. Ask several groups to report how their observations agree or differ with their expectations and assumptions. Discuss any lingering misconceptions, and summarize conclusions.

Day 3

In this session, students integrate the idea of force into their design considerations regarding elastic stretch. One way to make this transition is to use the following line of questioning (begin the class with a discussion about what causes the stretch in the rubber bands they have been using):

- *What causes the stretch and how does it happen?* (force attributable to the pull of gravity)
- *Can you think of a place where using applied mass would produce less stretch?* (the Moon, because its gravitational force is weaker [about one-sixth that of the Earth])

ACTIVITY 3 TEACHER PAGES

- *Can you think of a place where there would be no stretch at all?* (in orbit, where there is no effective gravitational pull)

Next write the equation $F = ma$ on the board to illustrate Newton's second law of motion. Ask students to consider how the masses (in kilograms) that they suspended can be equated to forces (in newtons). See if they remember that the acceleration of gravity is 9.8 meters per second squared (m/s^2) and know how this information can be used to calculate the force applied by a hanging mass. On the chart they made earlier, students recorded the number of units of applied mass and the corresponding distances the elastic stretched in each case. By multiplying each total applied mass (in kilograms) by 9.8 m/s^2, they can obtain the applied force (in newtons) for each case.

Have students describe what they have done in their own words, to assess whether they understand that they have quantified the force that a specified amount of hanging mass exerts (vertically) on their elastics. Do they further understand that their graphs now allow them to quantitatively describe the force they apply by hand, when the catapult is in operating position? Just by measuring *stretch,* they have force-calibrated the catapult. It is important for students to understand that it is the stretch in the rubber band that determines force, not the length of the band, which will not be zero when the force is zero. Furthermore, the rubber band need not be part of the catapult. If the catapult were cocked with a lever, a rubber band pulling on that lever could still be used to calibrate the force exerted at different angles of the lever.

As a further demonstration, you may wish to hook a spring scale calibrated in newtons onto a rubber band and pull it by hand for verification of the calibration.

To expand class involvement with science and math, you may wish to develop the concept that the area under each line in their graph represents the energy applied and stored. You might also wish to develop a classification of rubber bands by spring constants (k). Students can use Hooke's law ($F = kd$), to calculate k. This can be done algebraically from their data tables or by simply measuring slopes from their graphs. Students can discuss what higher and lower values for spring constants imply with respect to catapult performance.

The optional Projectile Motion student activity sheet (in Appendix A) addresses air friction as one of the factors in calculating projectile velocity. Other factors you and your students may wish to consider are humidity, temperature, projectile spin, density of the air at different altitudes, surface roughness and shape, velocity of the projectile, and aerodynamic lift.

Independent Science and Math Investigations

You may be interested in having your students conduct additional independent investigations of elastics as extensions of science and math applications. Here are some of questions they can pursue:

- *Compare the effect of using more than one rubber band (in parallel or series) versus using just one. Graph your results.*
- *Choose five different rubber bands and rank them from strongest to weakest. Graph your results.*
- *Choose five different rubber bands and rank them from most to least stretchy. Graph your results.*
- *Try some different configurations with your rubber bands (such as twisted, supported at greater or less initial tension, etc.) and compare the data you collect with that collected during class.*
- *Develop an explanation for what the area under the lines represents in your graph from the Investigating Elasticity homework assignment. Identify the relevant field of mathematics for quantifying the physical entity.*

Part II

3 DAYS

ACTIVITY 4 Development

OVERVIEW: DEVELOPMENT

You have completed basic experiments investigating elasticity. You have also reflected on the variables you need to control for improved catapult system and operator performance. To meet the Construct-a-Catapult design challenge, you will use your research findings to develop a prototype and create a calibrated, predictable firing guide for others to use. You will need to test-fire (and modify if necessary) your prototype until a consistent pattern—and statistical relationship—can be found between projectile landing distance and force applied. You are encouraged to expose your prototype to the scrutiny of other teams to supplement your internal evaluation and revision cycle.

Scope of Work

- Track and record milestones of your team's work in terms of the design process cycle.
- Build your prototype and test its performance.
- Measure and record projectile landing patterns correlating to forces applied.
- Modify design as necessary until your data are consistent enough for you to develop written instructions for a new user operating your catapult prototype.
- Exchange your prototype and operating instructions with another team to obtain peer feedback.

Evaluate your prototype critically and make modifications until you are satisfied with its improved performance, or are simply out of time. Remember to reflect on your process; how you go about your design and what you learn are key elements in the communication and assessment activities to follow. Good planning is essential to good design. Look again at the results of your research where you identified variables and wrote about the key differences in requirements for catapult performance. Brainstorm with your team to maximize identification of improvement alternatives—be creative. Review the Inquiry Process (p. 172) and Design Process (p. 176) activity sheets and be aware of where you are and where you are going in these iterative process cycles.

Parts of a Catapult

On the diagram below, sketch in and label a *stock*, *pedestal*, *slider*, and *trigger*. Refer to the Catapult Design History activity sheet for help.

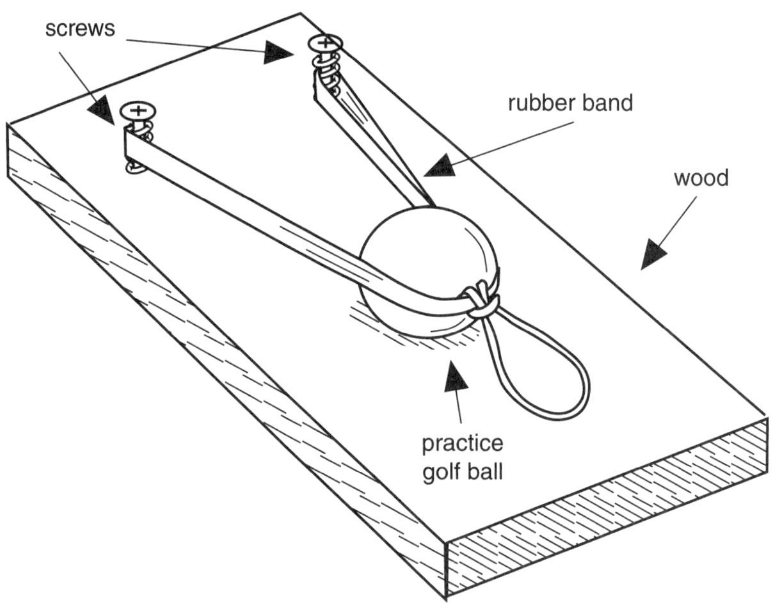

A Quick-Build Catapult

Part II

ACTIVITY 4
DEVELOPMENT

Resource List: Fasteners and Adhesives

In the lists below, circle the fasteners and adhesives you will use in building your catapult and write which parts of the catapult will be joined with those fasteners next to your choice. For example:

(Binder clips) —— *to attach rubber band retainer to stock*

Be prepared to explain your choices in terms of cost, speed, durability, and other factors.

Mechanical Fasteners

- Screws
- Nails
- Paper clips
- Binder clips
- Clamps
- Rivets
- Staples
- Bolts and nuts
- Hinges
- Wire
- String
- Elastic bands
- Clothespins
- Screw eyes
- Cup hooks
- Chain
- Pegs
- Slide/rail
- Clipboard

Adhesives

- Transparent tape
- Duct tape
- Masking tape
- Double-sided tape
- White glue (e.g., Elmer's)
- Paste/glue sticks
- Hot glue
- Construction adhesive
- Superglue
- Velcro strips and dots
- Adhesive foam mounting pads
- Contact cement
- Rubber cement

4

Beyond the Quick-Build: Second Pass

From your list in the previous activity, select the key variables you feel contributed most (positively or negatively) to the performance of your quick-build, and suggest what modifications you could make to meet the challenge goals. Record your current thinking in the following chart:

Most Significant Variables	Suggested Modifications

ACTIVITY 4 DEVELOPMENT

Team Feedback

Identify those variables that are interfering with the performance of the catapult. Make suggestions about what modifications could improve its performance. Record the variables and suggested modifications in the table below:

Interfering Variables	Suggested Modifications

4

SCILINKS® THE WORLD'S A CLICK AWAY

Topic: presenting data
Go to: *www.scilinks.org*
Code: CAC07

Making a Frequency Distribution

In the Challenge Event, you will exchange your catapult prototype and a user's manual with those of another team who will be asked to hit a target by following your instructions—without the benefit of practice or trial and error. For your team to succeed in meeting the Construct-a-Catapult design challenge, you need to make a launching graph that will guide your partner team to successfully operate your catapult. The launching graph must relate the *force* used to pull back the slider to the *distance* the projectile goes. By referring to the user's manual, the new team should be able to hit the target, no matter where it is placed.

You have probably seen variation in where the test projectiles land for a given force setting. You can refine your design and operating technique to reduce but not totally eliminate such variation. Statistical analysis of the variation will help you decide how to indicate and describe your catapult calibration markings for another user to interpret. The following activities will help you to find and describe patterns in your data variation. You can then use these patterns to make a launching graph that incorporates your calibration decisions.

First of all, you need to launch a number of projectiles at a variety of force settings and measure the distance they land from the catapult for each setting. Then make a series of frequency distributions so that you can analyze the spread pattern for your landing data. Refer to your frequency distributions and describe the overall patterns of projectile landings. Note and discuss anything unusual you observe about your data. Your frequency distribution will go into your user's manual.

ACTIVITY 4 DEVELOPMENT

Making a Launching Graph

Use the data from your frequency distributions to determine what distance, at each force setting, best represents the performance of your catapult for that force setting. Then, on a separate piece of paper, graph the relationship between amount of force (force setting) and distance traveled. Your launching graph will go into your user's manual.

Once you have made your graph, answer the following questions:

1. Describe the general shape of your graph.

2. What do the data tell you about the relationship between the amount of force used to launch a projectile and the distance the projectile travels?

3. Hooke's law states that stretch is proportional to the force applied. Does your graph indicate that distance traveled is also proportional to force applied?

4. What does your intuition, or knowledge of physics, suggest should be the relationship between force and distance traveled? (Hint: You may find the concept of energy conservation useful.)

Reflections On Your Design

1. What is the one feature of your catapult about which you are most proud?

2. Describe the concerns you have with the present design.

3. Describe in a paragraph or two the approach your team took in the designing and building of your catapult. Include a problem your team encountered with the performance of your catapult and how you went about solving it.

Part II

4 DAYS

ACTIVITY 4 Teacher Pages

OVERVIEW: DEVELOPMENT

As homework, students may read the Design Process optional activity sheet (included in Appendix A) and answer the design process questions to determine where in the design loop they've been, where they are now, and where they'll go next.

On the quick-build catapult diagram (p. 140) students superimpose and label additional performance enhancement parts referred to in the Catapult Design History activity sheet (p. 111). They then compare their answers with those of their teammates to develop a common vocabulary for discussing diverse design elements.

Students review a list of alternative fasteners and adhesives, make choices, and construct a working model of their design. Students unfamiliar with fastening technology will benefit if choices are made by team consensus. Students do a second pass on identifying variables impeding performance—this time of their redesigned launcher—and record their suggested modifications.

Each team pairs with another team, trades catapults for evaluation, and makes and receives suggestions for modifications.

Students launch projectiles with their prototype in a testing arena; they record landings, describe patterns, and construct a frequency distribution. Students use their judgment to select representative data to (1) graph for use as a launching guide and (2) use as a basis for final calibrations and operational instructions for their prototype catapult.

Students reflect on their designs, describing both strengths and weaknesses.

Preparation

- Prepare the fabrication zone: consider quantities of work space, materials, tools, and fasteners available, and optimize spatial arrangements for safe, efficient access.
- Set up the launch testing arena: If space in your room allows, give each group its own area to launch its catapult. If there is not enough space for this, designate a "launching arena" for groups to share.

- For easy cleanup and convenient data recording, you may want students to prepare the launching arena. Roll out 4–5 m of brown paper and tape it to the floor, making marks at meter intervals. Have colored sticky dots, cornstarch, or chalk dust available to mark the place where the projectile lands.
- Consider technological tools: If your students are proficient with computerized spreadsheet software or graphing calculators, the data they collect during test launches can be entered, manipulated, and printed. If they are not familiar with this technology, consider this an opportunity to demonstrate or expose them to the power of such tools.

Materials

For Each Student:

- Student Activity Sheets
    - » Beyond the Quick-Build (review)
    - » Overview: Development
    - » Parts of a Catapult
    - » Resource List: Fasteners and Adhesives
    - » Beyond the Quick-Build: Second Pass (assessment)
    - » Team Feedback Form
    - » Making a Frequency Distribution
    - » Making a Launching Graph
    - » Reflections on Your Design
    - » Design Process (optional)

ACTIVITY 4 TEACHER PAGES

For Each Team:

- Calibration Materials
    - roll of brown paper, 4–5 m in length
    - safety glasses or goggles
    - color-coding labels (assorted colors, round)
    - graph paper
    - metersticks
    - practice golf balls (or other lightweight projectiles)
    - chalk dust, cornstarch, or baking powder (for marking projectile landings)

For the Class:

- Redesign Supplies
    - screw eyes
    - screw assortment
    - rubber bands
    - binder clips of different sizes
    - bottle caps
    - hinges
    - corrugated board
    - spring and straight clothespins
    - lathe or wood molding strips

Time Requirement

The overall time constraint of four days requires efficient and productive team pacing to accomplish the design, construction, and testing of catapult prototypes with one or more iteration. Encourage students throughout this sequence to continue their team collaboration and to work intensively outside of class.

Teaching Suggestions

Student teams will vary in how long they take to complete different stages of the prototype process. Therefore, the timing and the order of activities described below should be flexible.

Discussion of Homework: Design Process (optional)

Discuss the homework in the Design Process activity sheet. A good initial question to pose might be: *Does the diagram depicting the design process accurately reflect your designing and building experience?* Other questions to ask from the homework include:

- *Which elements of this process have you already experienced?*
- *Where in the process do you think you are now?*

You may wish to make the criteria you will use to assess students' technological design capability available at this point and invite discussion.

Suggest to students that if they become stuck at any point, they should refer to their portfolio notebook for ideas about what to do next. Highlight, or elicit, examples of how students already used or might soon use the design process loop in nonlinear progression.

Starting to Redesign the Catapult

Students may choose to proceed with the redesign of their catapults either by improving on their quick-builds to make them more accurate or by starting fresh. In either case, they should (a) preserve or duplicate the elastic system they worked with in the research phase or (b) recalibrate a newly designed system. Advise them also to keep the users of their catapults in mind. Remind them that teams will be *exchanging* catapults to meet the final challenge. Therefore, students must make a catapult and instructions for operating it written in such a way that other students can operate the catapult easily and successfully.

Indicate the range of distances to target that may be employed in the final challenge, stressing that the exact distance will not be announced before the challenge. Operating instructions for the catapult must inform the user how to hit a target on the first attempt at any distance within the given range.

ACTIVITY 4
TEACHER PAGES

Resources for Redesigning

Before students begin redesigning, have them complete and discuss the Parts of a Catapult activity sheet—first individually, then as a team—to define the team's design vocabulary. They could also review the following resources to help them get started:

- Beyond the Quick-Build: This activity sheet, which students did in Activity 2 (p. 120), can be used to refresh their memories about what kinds of modifications they need to make as they redesign their launchers.
- Catapult Design History: This sheet includes diagrams and descriptions of types and parts of catapult stocks, pedestals, sliders, triggers, and propulsion systems.
- Resource List: Fasteners and Adhesives: This worksheet lists possible ways to fasten parts of their prototype together.

Safety Alert!

Remind all students in the launching and landing zones to wear safety goggles or glasses.

Timely Feedback

As students move beyond their quick-builds, it is essential that teams be allowed to grapple with the challenges of designing, building, and testing their catapults—and of revisiting these processes. But it is also important that they receive input at critical times. One source of input is the team itself or perhaps a team launching nearby. Another important source is timely intervention from you. Very often, a suggestion about how to solve a problem, or a reference to a diagram or picture, will stimulate action. Just be careful to avoid intervening too early and too often.

Evaluation and Reflection

After they have had a chance to struggle with the redesign of their catapults, or have become stuck, students have the opportunity to evaluate and reflect on their redesign using the activity sheet Beyond the Quick-Build: Second Pass. You may wish to use this activity sheet as a way of assessing students' growing ability to identify relevant variables. You might also assess their troubleshooting or problem-solving ability, as revealed in modifications they have made to address problems they have encountered.

Following this reflection/assessment, pair each team with another for the purpose of providing feedback about each other's catapults. Suggest to partner teams that they demonstrate their catapults to each other and that each team collaborate to provide written feedback to the other on the *Team Feedback Form*. The goal here is twofold: for students to share ideas and for them to obtain helpful feedback from others.

Both the second pass and the feedback forms can be used more than once.

Once groups have consulted and provided feedback to each other, provide some time and space for each team to discuss and decide next steps.

Making a Force Scale

If students have not yet made markings for a force scale on their catapult, they have probably been using trial and error to hit their target. The force scale's purpose is to indicate as precisely as possible how far back the rubber band or slider should be pulled to apply specific force. Correlating landing distance with applied force is necessary for achieving repeatable results. Students should refer to the table they made in the research phase to relate stretch distance to force applied. They will most likely inscribe or attach the force scale on their sliders or stocks. To prevent a user from exceeding the limit of elasticity for their device (which would deform the elastic and render their markings useless), teams should consider making a slider stop or restraint.

Some catapult designs do not allow for a simple stretch scale along the slider. The trebuchet, for example, has no slider and is cocked with a rotating lever. Calibration can still be accomplished, however, using the same principles: Use a calibrated rubber band to pull the rotating lever to a particular position. The stretch of the rubber band is still a measure of the force needed to reach that position.

Collecting Data

To test their catapults, students launch balls into the launching arena using a fixed launch angle and a range of force settings. As each projectile is launched, it is helpful if one team member observes where it lands and puts a colored dot coded for the force setting at the landing place. Rolling the projectile into a powder such as colored chalk, cornstarch, or baking powder ahead of time makes the spot where it lands more visible.

ACTIVITY 4 TEACHER PAGES

Deciding How Good Is "Good Enough"

Once groups start testing the accuracy of their catapults, it is important to discuss with them how accurate their catapults need to be. The answer to this question will depend on your own goals and what is realistic, given the materials students are working with and their skills as problem solvers, designers, and builders. A suitable range of accuracy might be for the majority of balls launched at a given force setting to land at distances that are within 10–20% of each other. Alternatively, you might create a Challenge Event goal to land within a target zone (for example a tin pie plate) three out of five times.

Students will notice that even when they pull the slider back to the same place each time, the projectile does not always land in the same place; at best, the data will cluster within a certain range. Discuss with students ahead of time how many trials they think will give them a clear idea of the catapult's performance variation at that setting. This will depend on the accuracy that has been selected. In most cases, about 10 projectiles at each force setting should be adequate.

Whatever the standards you set for your students (or they set for themselves), remind them of the need to analyze and improve both the catapult system design and the operator launching techniques to achieve consistent performance. As your students are testing their catapults, look for evidence that they are evaluating and using their findings to make both system and operating improvement decisions.

Representing Data in a Frequency Distribution

To obtain a clear picture of their catapult's performance over a range of force settings and to show the spread of the data (variation or "error") at each of these force settings, students mark the landing points with dots or labels coded according to force setting, measure the distance each ball lands from the catapult, and represent these data graphically. As a visual record, they may wish to photograph the arena to show the overall landing patterns, draw an arena diagram, or, if paper was used on the arena floor, simply save the marked paper.

One way for them to represent patterns in their data quantitatively is with a frequency distribution. If your students have not had much experience describing data statistically, you may want to put a sample distribution on the board and discuss as a class how to find which distance is most representative at a given force setting.

Students will display the patterns in their data on the Making a Frequency Distribution activity sheet.

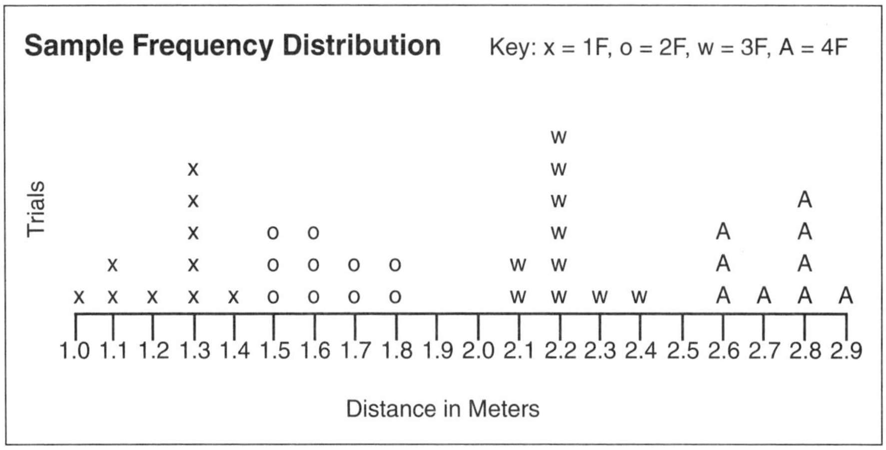

Representing Data in a Line Graph

Students study the frequency distributions they made and determine a single distance that best represents the relationship of launch force to distance traveled for each data cluster. For a certain force setting, students may notice a mode or modal cluster in the data pattern and choose that distance as the most representative. In another situation, there may be no clear mode, and students may decide that the average best represents the performance of the catapult for that force setting. With such summary statistics of their data, students make a simple, clear launching graph for use in the Challenge Event, keeping in mind that others will have to understand it.

Each team makes their custom launching graph on separate paper as part of the Making a Launching Graph activity sheet. This graph will be included later in their user's manual. Addition of grid lines, error bars, and other enhancements should be

ACTIVITY 4
TEACHER PAGES

reviewed by teams to describe whether such enhancements would assist or confuse the prospective user in the Challenge Event.

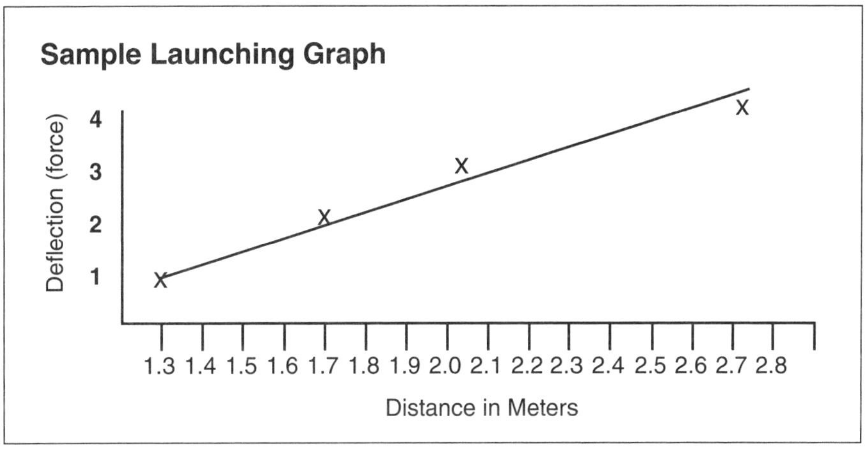

Optional Extension: Projectile Motion

Students might compare their data with expectations from textbook descriptions and mathematical models of the parabolic paths of projectiles. They can determine the mass of the ball, apply known forces, and observe or even videotape trajectories and record landings. Most models apply to cases that have been simplified to ignore complicating effects of air friction. The projectiles used here, however, are acceptable in the classroom because they are specifically designed with low mass relative to surface area to capitalize on air friction and limit velocity and distance traveled. For these reasons, one cannot expect data from the catapults to follow frictionless models precisely.

By making comparisons of standard assumptions to real circumstances, students may better understand why different statistical summaries fit different force settings. This could lead them to create custom corrective factors for their launching graphs that are rooted in new conceptual understanding.

Design Self-Assessment

As a final development activity, students reflect on what they have done thus far in Reflections on Your Design. They are asked to identify positives and negatives about their current design and to describe briefly their team approach and process. This activity sheet is short, and you can add your own questions.

Assessing Students' Design and Build Capabilities

There are several key elements of students' design and build capabilities you could assess throughout the development phase:

- *How well students are able to develop solutions.* Developing ideas through to workable solutions is at the core of technological design. Look for evidence of students' ability to do this, both in your ongoing observations of students at work and in the responses they give on the activity sheet Beyond the Quick-Build: Second Pass.
- *How well students are able to evaluate the processes they have used.* This includes the extent to which they are able to identify strengths and weaknesses of their catapults. Look for evidence of students' ability to do this, both in your ongoing observations of students at work and in the responses they give on the activity sheet Reflections on Your Design.
- *To what extent do students exhibit ownership of the task?* Did it change with time? How much initiative did they take? Look for evidence of this in your ongoing observations of your students.

ACTIVITY 4
TEACHER PAGES

Sample Student Catapult Designs

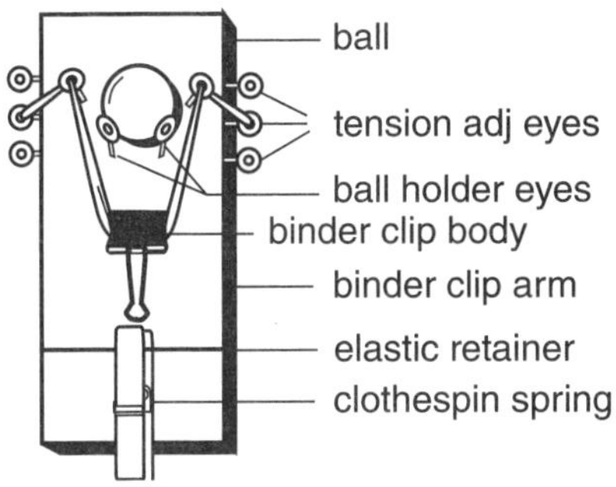

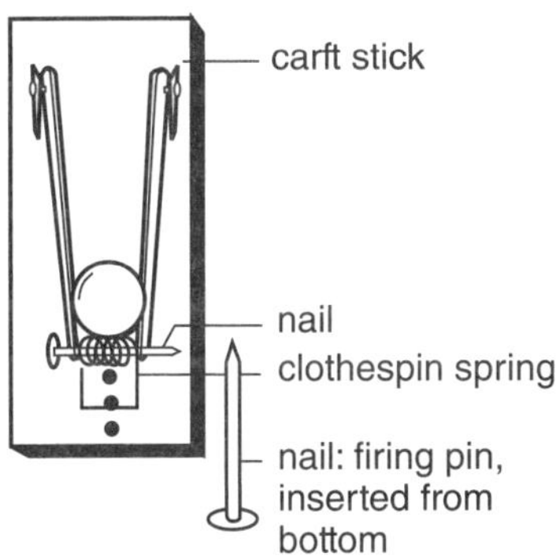

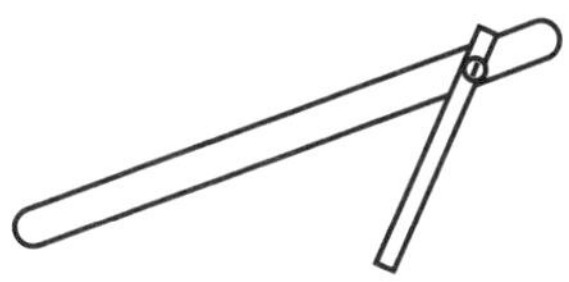

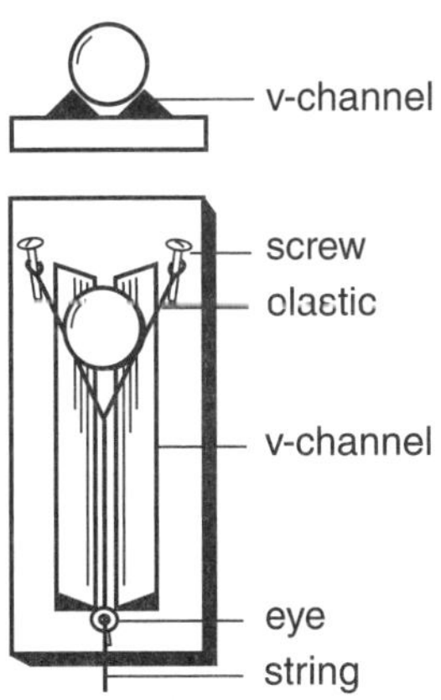

v-channel variations

(continued)

Sample Student Catapult Designs *(continued)*

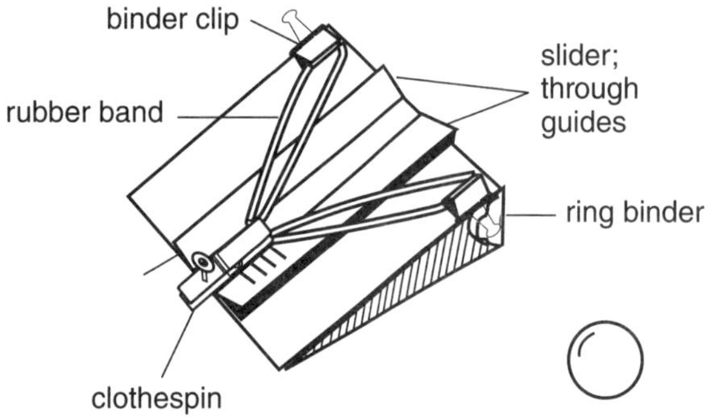

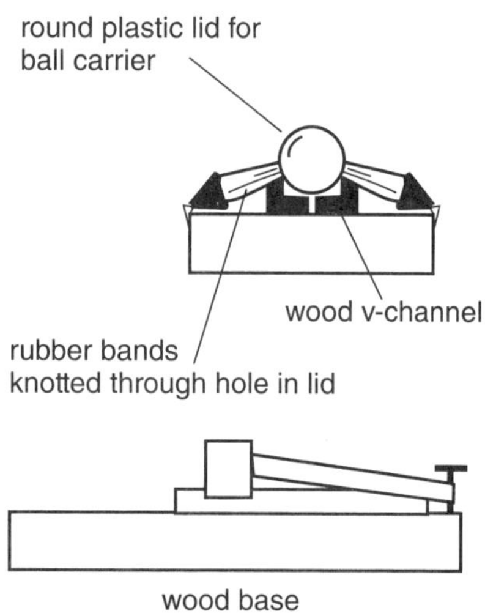

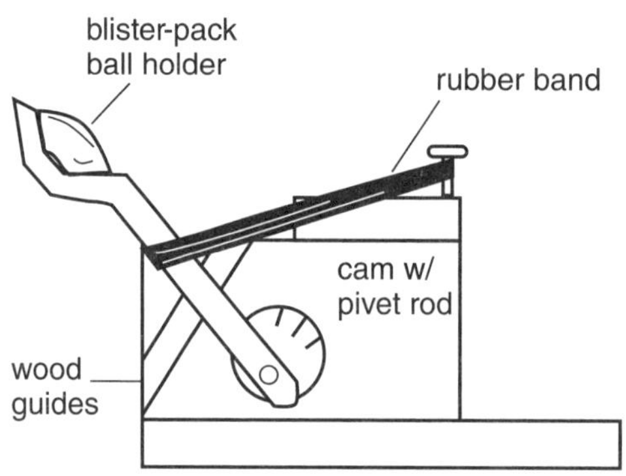

Part II

3 DAYS

ACTIVITY 5 Communication

OVERVIEW: COMMUNICATION

Your research and development team has made a prototype ready for demonstration in the final catapult Challenge Event. Now another team—using your draft written instructions—will operate your catapult in the event. For this reason, your team must communicate clear, effective operating instructions in a user's manual that also summarizes and documents your efforts in constructing your catapult.

Think about the discoveries you have made while working through Construct-a-Catapult. Communicate the important science and technology topics that relate to your effort, and describe your work in a way that allows a prospective user to understand the principles involved.

Scope of Work

- Demonstrate the calibration accuracy of your prototype according to Challenge Event rules.
- As a team, write a user's manual with instructions for others to successfully operate your catapult within performance specifications.
- Assess your own learning by comparing your postproject Snapshot of Understanding answers with those you wrote at the beginning of the unit.
- Submit your portfolio for individual assessment.

5

Creating a User's Manual

The final activity is for you to communicate the results of your work in a user's manual. Your goal is to include information that would interest others in using your catapult and inform them about its capabilities.

An important skill to demonstrate in your user's manual is the ability to communicate clearly about your catapult in writing. The writing in each section should be well organized and clear enough for someone unfamiliar with your team's catapult to understand.

You may find it interesting to consult several real user's manuals before you begin, but don't feel that yours needs to be just like these. Any major appliance will have a user's manual, as will computer software or an electronic device. Have fun thinking about your own ways of describing your catapult.

User's Manual Table of Contents

Consult the table of contents shown below for a list of the topics you should address in your user's manual. Some of these topics, like the firing tips and launching graph, have already been done. Others, like the mission statement and scientific principles of operation, still need to be written.

Table of Contents

I. Team Mission Statement

II. System Overview
   A. Parts and Materials Specifications
   B. Performance Specifications

III. Operating Instructions
   A. Safety Measures
   B. Firing Tips
   C. Launching Graph

IV. The Science in Our Catapult

ACTIVITY 5
COMMUNICATION

V. Appendixes (optional)
   A. Similar Product Comparisons
   B. Preventive Maintenance
   C. Repair and Recalibration
   D. Disclaimers and Warranty

Members of your team should divide up the responsibility for writing each of the sections. Before you do so, read the suggestions below to obtain a clear idea of what each section is about.

Suggestions for User's Manual Sections

- **Team Mission Statement:** State the purpose for your catapult. Take into consideration what your catapult's capabilities are as you think about what uses it could have. The mission statement should be a total team effort. First brainstorm as a team, then assign the actual writing to one member.
- **System Overview:** This section includes a sketch of your catapult with labeled parts.
- **Parts and Materials Specifications:** This section lists details such as size, shape, composition, and quality of each component part.
- **Performance Specifications:** This section is the place to describe the capabilities of your product. What are the minimum and maximum distances your catapult can handle? How accurate is your catapult? (Your frequency distributions provide information about accuracy.)
- **Operating Instructions:** This section should provide a step-by-step set of instructions for operating your catapult.
- **Safety Measures:** This section describes what precautions the user needs to take when using the catapult.
- **Firing Tips:** This section summarizes your advice ("do's and don'ts") to the user. Provide as many tips as you can for helping the user get accurate and reliable performance.

- **Launching Graph:** In this section you should modify or explain how to use your existing launching guide in the context of your catapult's newly determined mission (purpose or intended use).
- **The Science in Our Catapult:** Here you should discuss the scientific principles involved in the operation and use of your device. Consult and include the appropriate student activity sheets to address this topic.
- **Appendixes are optional.** Use them if you wish to include any of the following additional information:
    - » For *Product Comparisons*, describe each team's model in terms of its strengths and weaknesses. Think of yourself as a member of a large catapult manufacturer and your team's catapult as one in a range of choices for users with differing needs.
    - » For *Preventive Maintenance*, you might let the user know what kind of ongoing maintenance you think your catapult will need to continue performing well.
    - » *Repair* directions can address how to fix or rebuild the different parts when they break down.
    - » *Recalibration* directions could describe how to recalibrate your catapult's slider system if the elastic breaks or if the user wants to change the kind of elastic he or she is using.
    - » *Disclaimers and Warranty* addresses which parts your team as manufacturer will be responsible for should some repair be necessary and which parts you will not stand behind.

ACTIVITY 5
COMMUNICATION

The Challenge Event: Usability Testing

To demonstrate that you have met the Construct-a-Catapult design challenge, you will put the prototype catapult you have designed and built to the ultimate test—having someone else operate it according to your instructions. This is comparable to "usability testing" in a real product development cycle. Similarly, you will be the user/tester of a partner team's unit.

Your teacher will provide details of the event structure and procedure. Ingredients for success may include:

- a clear and accurate user's manual
- simple, understandable instruction (launching tips)
- durable construction
- well-tested, precise calibration clear to any user

Notes

Use the space below to record your teacher's detailed instructions and criteria for success for the Challenge Event, as well as your own thoughts and reminders both before and after the event.

5

SNAPSHOT OF UNDERSTANDING
What I Now Know About Catapults

1. a. What variables are important to control for catapult accuracy?

   b. What scientific principle(s) can you use to describe how the catapult you designed and built operates?

2. What are the main features of a catapult? Make a sketch and label its parts, or describe.

3. List current uses of catapult-like devices of which you are aware.

ACTIVITY 5
COMMUNICATION

4. Describe how you went about designing and building your catapult.

5. What do the following terms mean?

force:

energy:

ACTIVITY 5 Teacher Pages

3 DAYS

OVERVIEW: COMMUNICATION

Teams summarize what they have learned about their catapult by creating a user's manual that includes information about how to use their catapult, how it is constructed, its principles of operation, and performance specifications. A sample of topics for students to address, along with explanatory text, is provided on the Creating a User's Manual activity sheet. In the Challenge Event, teams exchange catapults and launching graphs, with associated instructions and firing tips. Each team operates another team's prototype. The most successful design is the one that can be used most accurately by another team.

As a final assessment, students answer questions similar to those at the beginning of the unit in the final Snapshot of Understanding.

Preparation

- Prepare a Challenge Event plan, location, and procedure.
- Consider offering spectator invitations, a competitive aspect, and a focus on launching guides.
- Provide (or have students bring in) a few different examples of user's manuals.
- Customize the user's manual Table of Contents to fit the time and educational objectives for your class (or encourage students to make such decisions).
- Prepare a grading plan for your evaluation of the team and individual effort.
- Arrange for use of word processing/graphics (or CAD) computer stations (optional).
- Arrange flexible seating to facilitate team discussions and writing (optional).

ACTIVITY 5
TEACHER PAGES

Materials

For Each Student:

- Student Activity Sheets
  - » Overview: Communication
  - » Creating a User's Manual
  - » The Challenge Event: Usability Testing
  - » Snapshot of Understanding

For Each Team:

- safety glasses or goggles
- target for the Challenge Event
- word processing or graphics stations as available (optional)

Time Requirement

This unit can be conducted in three class sessions: one for writing the team user's manual (plus homework), one for the Challenge Event, and one for reflection and final self-assessment (allow about 20 minutes for completion of the Snapshot of Understanding).

Teaching Suggestions

Creating the User's Manual

Discuss with your students which topics to include in their user's manuals and suggest that they develop a plan for delegating parts of the writing within their teams. Also suggest that they take into consideration what the strengths and weaknesses of their catapult may be compared with the catapults of other teams. This may help them consider how to write instructions that can be used by students who are used to another kind of catapult.

Encourage teams to test drafts of their user's manual with people from the other teams rather than with members of their own team. At the same time, within each team, every person should review the other members' sections before the report is assembled.

If time permits, include a brainstorming session for possible uses or markets for the catapults. Examples could be delivering wastepaper to the circular file, passing cherry tomatoes at the dinner table, or providing an automatic partner for Ping-Pong practice.

During the Challenge Event, the ability to communicate clearly in writing via a user's manual may prove to be more important than the design and construction of the catapult. Remind students that whether in science or technology, clear communication of purpose, method, and process is not only required, it is often the most important element of success in the marketplace.

Safety Alert!

Remind all students to wear safety glasses or goggles, particularly in the launching or landing zones.

Meeting the Construct-a-Catapult Challenge

When groups have completed their user's manuals, it is time to conduct the Challenge Event. Arrange a system for groups to exchange catapults. To avoid advance coaching, do not tell teams ahead of time which team gets which catapult. The target range you select should also be unpredictable.

To choose a target distance, you may wish to examine students' data to find a target distance that lies between data points already measured. Alternatively, you might select outside their data range, requiring students to interpolate or extrapolate.

Make sure that accurate data are maintained on every team's operation of the catapult so that fair comparisons can be made and the performance of each catapult can be ranked.

If time permits after the Challenge Event, consider the following questions in a class discussion:

- Which catapult had the greatest variation in data? Which characteristics of this catapult could account for this variation?
- Which catapult demonstrated the greatest accuracy? Which characteristics of this catapult could account for its accuracy?
- Did each team operate the catapult according to instructions? If not, were the user operating procedures at fault, or did the team misread them?

ACTIVITY 5 TEACHER PAGES

Continue the discussion in the following class period, leaving just enough time for students to complete the Snapshot of Understanding.

Completing the Snapshot of Understanding

After students complete the final Snapshot of Understanding, provide a brief period of time for them to compare their new answers with those on their preunit Snapshot. (Hand the first Snapshot back at this time.) It can be very empowering for students to see for themselves how much they have learned.

APPENDIX A
Side Roads

The material in this appendix is intended to support activities that you may choose to add to those described in the core unit. The following pages correspond to the side road suggestions made in each activity. Many of these are key activities, but they have been placed in this appendix because they can fit in several different places in Construct-a-Catapult—exactly where they are used is a matter best decided by you in response to student questions and feedback. Some activities may be profitably used more than once. An analysis of the design process, for example, will provide different insights when used in the research activity than when used in the design activities.

In this appendix:

- Homework: Individual Information Search
- Inquiry Process
- Projectile Motion
- Design Process
- Additional Suggestions

Appendix A
SIDE ROADS

HOMEWORK
Individual Information Search

The introduction to this unit on catapults gives you a sense of the variety of forms and uses of the machine. The pictures included show three different types of catapult: the ballista, the onager, and the trebuchet. Select one of the three types of catapults and perform an in-depth information search. Use the questions outlined below as a guide, and put together a report for the class that includes illustrations, written materials, and a list of relevant references for those who wish to explore your topic further.

I. The history of your type of catapult:
   A. For what purpose was this type of catapult used or designed?
   B. Did it persist in use for the original purpose?
   C. How did its uses and/or purposes change through history?
   D. Are there contemporary uses of the device itself, or as part of another machine? Give examples.

II. The design of your type of catapult:
   A. Describe the original design. Use illustrations to identify parts and their functions, and materials used.
   B. Was the original design modified throughout history? How and why?
   C. Has the original catapult of this type been supplanted by some other machine? What and why?
   D. Are there other machines that incorporate this device? Give examples.

III. The science and technology in your catapult:
   A. What scientific principles or disciplines contributed to the design?
   B. What technologies were applied to create this type of catapult?

IV. References you used and recommend:
   A. Books
   B. Articles
   C. Films, videos
   D. Websites

A

Inquiry Process

The inquiry process is often viewed as a cycle of action that repeats until the investigators reach a satisfying solution. It can be described with seven basic elements:

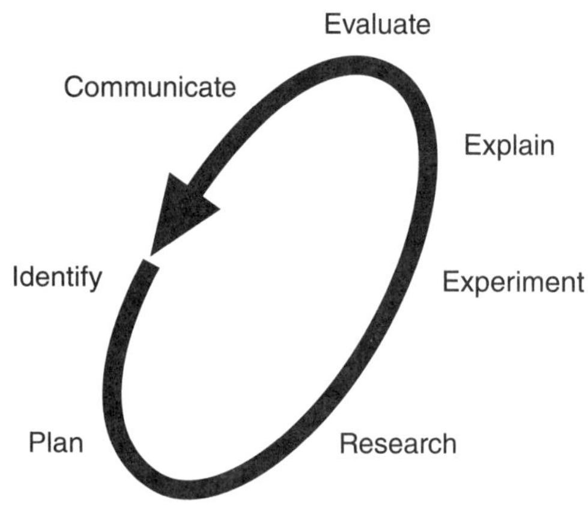

- *Identify* and clarify questions. Understand the issue or problem, and make a testable hypothesis.
- *Plan* appropriate procedures. Brainstorm, draw and write ideas, clarify their ideas, and suggest possible strategies or methods.
- *Research* major concepts. Learn what is known about the situation from sources other than actual investigation, and obtain information from preliminary experiments. Decide what technology, approach, equipment, and safety precautions are useful. Document your experiments and log your data.
- *Experiment.* Use tools and measuring devices to conduct experiments. Use calculators and computers to store and present data.
- *Explain* logical connections. Analyze your data. Formulate explanations using logic and evidence, and possibly by constructing a physical, conceptual, or mathematical model.
- *Evaluate* alternatives. Compare your explanations to current scientific understanding and other plausible models. Identify what needs to be revised, and find the preferred solution.
- *Communicate* new knowledge and methods. Communicate results of your inquiry to your peers and others in the community. Construct a reasoned argument through writing, drawings, and oral presentations. Respond appropriately to critical comments.

Appendix A
SIDE ROADS

Questions

Read the following questions, but do not answer them until after your team has experienced working together on the design challenge research activities.

1. Make your own checklist of team activities that correspond to steps in the cycle described on the previous page.

SCILINKS® THE WORLD'S A CLICK AWAY

Topic: scientific inquiry
Go to: *www.scilinks.org*
Code: CAC08

2. Create your own version of the inquiry process using words and pathways that fit your team's activity.

3. What shape is your inquiry pathway diagram (circle, spiral, cascade, other)?

4. How and where do the seven steps described on page 172 fit within your process description?

A

Historical Note

The ENIAC computer was created in the mid-1940s, using approximately 18,000 vacuum tubes, at a cost of $500,000 for military purposes such as calculating ballistic trajectories.

Projectile Motion

A projectile is any object that has been launched into the air and moves freely under only the influences of gravity and air resistance. The effect of gravity is a constant downward acceleration from the moment of launching, usually specified with the value $g = 9.81\ m/s^2$. The effect of air resistance is often relatively small and may be difficult to predict as a combination of object surface drag, air speed, and direction.

Launch Angle Effect

The angle of launch (Θ) is significant because angles above the horizontal add an upward component of velocity against the acceleration of gravity, increasing the time of flight (but not necessarily the horizontal distance traveled) before landing. A typical two-dimensional graph of projectile motion (here using $\Theta = 45°$ and v_0 [initial velocity] = 5 m/s) is a parabola:

$$y = x \tan 45 - \frac{9.81}{2 \times 5^2 \cos^2 45} x^2$$

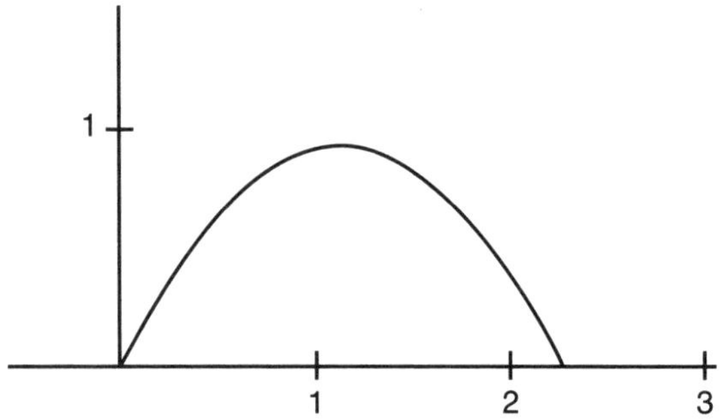

On the graph above, superimpose sketches of the trajectory curves you would expect for similar projectiles launched with equal force, but at angles of 60° and 30°. At what launch angle(s) do you expect to obtain (a) greatest horizontal distance and (b) greatest vertical height?

APPENDIX A
SIDE ROADS

Air Resistance Effect

When precise prediction is not required, air resistance is often not calculated. In the Construct-a-Catapult design challenge, air resistance is an important factor in the behavior of practice golf balls used as projectiles. A key design criterion for practice golf balls is to limit the distance of travel without skewing direction. This performance characteristic is useful for the scale of construction and testing possible within the classroom.

Research air resistance effects on projectiles and list the factors you discover in the space below. Seek formulas for mathematical expression of air resistance and describe how to experimentally quantify a "damping coefficient" for your projectile.

Initial Velocity Effect

Refer to your launching graph and use your prior knowledge about Newton's second law of motion ($F = ma$) and your findings about air resistance to explain any changes observed in the relationship between applied force and distance at increasing velocities.

SCILINKS
THE WORLD'S A CLICK AWAY

Topic: force
Go to: *www.scilinks.org*
Code: CAC01

Topic: energy
Go to: *www.scilinks.org*
Code: CAC02

Topic: elasticity
Go to: *www.scilinks.org*
Code: CAC03

Topic: gravity
Go to: *www.scilinks.org*
Code: CAC05

Topic: friction
Go to: *www.scilinks.org*
Code: CAC09

A

Design Process

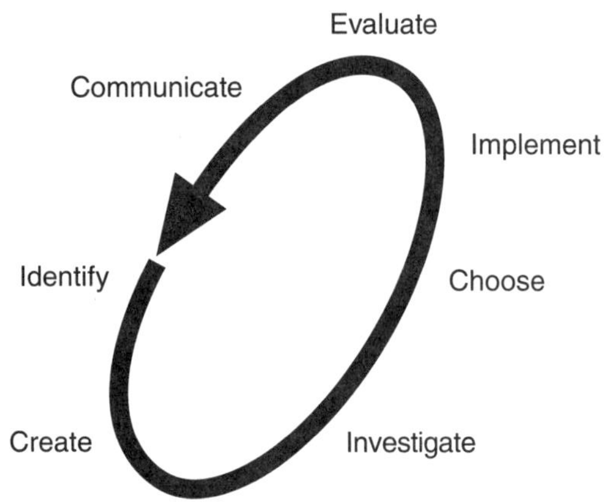

The design process is often viewed as a cycle of action that repeats until the designers reach a satisfying solution. It can be described with seven basic elements: identify, create, investigate, choose, implement, evaluate, and communicate.

- *Identify* and clarify the situation. Understand the challenge or problem, including the criteria for success and constraints on the design.
- *Create* solutions. Brainstorm, draw and write ideas, and suggest possible strategies or methods.
- *Investigate* possibilities. Learn what is known about the situation, and what technology or approach could be useful. Conduct experiments to test your ideas.
- *Choose* a solution. List the solutions most likely to be successful, and make decisions for how well each solution meets the design challenge or solves the problem.
- *Implement* the design. Learn that a successful design often depends on good fabrication, whether it is a scaled or life-size version of the product.
- *Evaluate* the design. Perform tests to obtain the feedback that informs them about the parts of the design that worked or needed improvement.
- *Communicate* the solution. Present your designs to your peers and others in the community, communicating your ideas through drawings, writing, formal presentations, informal discussions.

APPENDIX A
SIDE ROADS

Questions

After reading about the design process, answer the following questions:

1. Which elements of the process have you already experienced?

2. Which elements have you not yet experienced?

3. Where in the process do you think you are now?

4. What will your next steps be?

Additional Suggestions

If you are comfortable with calculus, you may want to point out that energy is calculated by the area under a force versus stretch graph. In the case of Hooke's law, the graph will be a straight line of slope k. The area under the graph will be a triangle, and the equation $E = (½)kS^2$ is the area of a triangle of base S and height kS.

You should be well prepared for the likelihood that students will have lots of questions, such as "What triangle produces $(½)mv^2$?"

Part II

APPENDIX B
Text Reconstruction Exercises

Force

The paragraphs below describe how the concept of force can help you better understand the workings of your catapult. To prevent this information from falling into the wrong hands, the order of the sentences in each paragraph has been jumbled. Your task is to rearrange the sentences in their correct order.

¶2

___ The reason it is difficult is that it is similar but not entirely the same as the concept we use in everyday speech.

_1_ The concept of force as it is used in physics is both simple and, at the same time, very difficult to understand.

___ The best advice we can give is to look for these differences.

___ Gradually you will gain enough experience to understand the physics way of talking.

¶2

_2_ His second law, $F = ma$, states that forces affect not the speed of an object but rather the rate of change of speed.

___ Isaac Newton became famous by discovering how force was related to motion.

___ A constant force will make an object go faster and faster and faster ….

¶3

___ There are two kinds of force you need to know how to calculate for designing a catapult.

___ The second force is gravity.

_4_ It is given by the equation $F = mg$, where g is the acceleration of gravity (on the surface of the Earth that is 9.8 meters/[second]^2)and m is the mass being pulled by gravity.

___ The force of pull of a stretched rubber band is given by Hooke's law: $F = kS$, where S is the amount of stretch of the rubber band.

B

Elasticity

The paragraphs below describe how the concept of elasticity can help you better understand the workings of your catapult. To prevent this information from falling into the wrong hands, the order of the sentences in each paragraph has been jumbled. Your task is to rearrange the sentences in their correct order.

¶ 1

___ There are deformations beyond which there is no return.

___ A rubber ball is a good example of a highly elastic object; a windowpane is a good example of a less elastic object.

_2_ Everything is elastic to some extent, but every object also has limits.

___ The elastic property of matter is its ability to deform under forces and to return to its original shape when the forces are removed.

¶ 2

___ You push twice as hard and you will get twice the deformation.

___ In this case, the idea that force (F) is proportional to stretch (S) can be expressed as $F = kS$.

_3_ This is most easily seen in a rubber band, which will stretch some amount for 1 unit of force and twice as far for 2 units of force.

___ Hooke's law states that the region of elasticity is governed by a rule in which the amount of deformation is proportional to the applied force.

___ When two variables are directly proportional, they always follow one very simple mathematical equation of the form $A = kB$.

APPENDIX B
TEXT RECONSTRUCTION EXERCISES

Energy

The paragraphs below describe how the concept of energy can help you better understand the workings of your catapult. To prevent this information from falling into the wrong hands, the order of the sentences in each paragraph has been jumbled. Your task is to rearrange the sentences in their correct order.

¶ 1

_4_ As long as the catapult is cocked, this energy remains hidden in the elastic (physicists call this hidden energy "potential").

____ In this equation, E stands for energy, k the elastic constant (different for each rubber band), and S is the amount of stretch.

____ The energy stored in an elastic medium (a rubber band in our case) can be calculated from the formula $E = (½)kS^2$.

____ When you fire a spring-loaded catapult, you convert elastic potential energy to kinetic energy.

¶ 2

____ If we assume all the energy goes into the projectile, we can equate the two energy equations and write: $(½)mv2 = (½)\ kS^2$.

_5_ In this equation, E stands for energy, m for mass, and v for velocity.

____ After the catapult fires, the energy appears in the projectile as "kinetic" energy.

____ Some energy disappears in friction and recoil, but we usually ignore these factors.

____ The kinetic energy of any moving object can be calculated from the equation $E = (½)mv^2$.

____ The catapult in which friction and recoil can realistically be ignored is called an ideal catapult.

B

SCILINKS
THE WORLD'S A CLICK AWAY

Topic: energy
Go to: *www.scilinks.org*
Code: CAC02

Topic: elasticity
Go to: *www.scilinks.org*
Code: CAC03

Topic: gravity
Go to: *www.scilinks.org*
Code: CAC05

Topic: mass
Go to: *www.scilinks.org*
Code: CAC06

Topic: friction
Go to: *www.scilinks.org*
Code: CAC09

Text Reconstruction Key

The paragraphs below show the correct order of sentences for each text reconstruction exercise. When you hand out the initial homework assignment, ask students to number the sentences in each paragraph so as to put them in the correct order. It is also highly beneficial to ask students to rewrite the paragraphs once they have determined the correct order.

Force

Paragraph 1

1. The concept of force as it is used in physics is both simple and, at the same time, very difficult to understand.
2. The reason it is difficult is that it is similar but not entirely the same as the concept we use in everyday speech.
3. The best advice we can give is to look for these differences.
4. Gradually you will gain enough experience to understand the physics way of talking.

Paragraph 2

1. Isaac Newton became famous by discovering how force was related to motion.
2. His second law, $F = ma$, states that forces affect not the speed of an object but rather the rate of change of speed.
3. A constant force will make an object go faster and faster and faster ….

Paragraph 3

1. There are two kinds of force you need to know how to calculate for designing a catapult.
2. The force of pull of a stretched rubber band is given by Hooke's law: $F = kS$, where S is the amount of stretch of the rubber band.
3. The second force is gravity.
4. It is given by the equation $F = mg$, where g is the acceleration of gravity (on

APPENDIX B
TEXT RECONSTRUCTION EXERCISES

the surface of the Earth that is 9.8 meters/[second]^2)and *m* is the mass being pulled by gravity.

Elasticity

Paragraph 1

1. The elastic property of matter is its ability to deform under forces and to return to its original shape when the forces are removed.
2. Everything is elastic to some extent, but every object also has limits.
3. There are deformations beyond which there is no return.
4. A rubber ball is a good example of a highly elastic object; a windowpane is a good example of a less elastic object.

Paragraph 2

1. Hooke's law states that the region of elasticity is governed by a rule in which the amount of deformation is proportional to the applied force.
2. You push twice as hard and you will get twice the deformation.
3. This is most easily seen in a rubber band, which will stretch some amount for 1 unit of force and twice as far for 2 units of force.
4. When two variables are directly proportional, they always follow one very simple mathematical equation of the form $A = kB$.
5. In this case, the idea that force (F) is proportional to stretch (S) can be expressed as $F = kS$.

B

Energy

Paragraph 1

1. When you fire a spring-loaded catapult, you convert elastic potential energy to kinetic energy.
2. The energy stored in an elastic medium (a rubber band in our case) can be calculated from the formula $E = (½)kS^2$.
3. In this equation, E stands for energy, k the elastic constant (different for each rubber band), and S is the amount of stretch.
4. As long as the catapult is cocked, this energy remains hidden in the elastic (physicists call this hidden energy "potential").

Paragraph 2

1. After the catapult fires, the energy appears in the projectile as "kinetic" energy.
2. Some energy disappears in friction and recoil, but we usually ignore these factors.
3. The catapult in which friction and recoil can realistically be ignored is called an ideal catapult.
4. The kinetic energy of any moving object can be calculated from the equation $E = (½)mv^2$.
5. In this equation, E stands for energy, m for mass, and v for velocity.
6. If we assume all the energy goes into the projectile, we can equate the two energy equations and write: $(½)mv^2 = (½)kS^2$.

APPENDIX C
Sample Answers

Identifying Variables

Activity 2, p. 119

Parts of the Catapult System	
Variable	**Range of Variation**
Example: *Angle of the catapult*	*Flat, between 0° and 45°, between 45° and 90°*
Stretch of elastic	*A little—really breaks*
Constrain projectile to center of run	*No adjustment available*
User Operating Procedures	
Variable	**Range of Variation**
Example: *How you position the rubber band*	*high on the nails, in the middle of the nails, low on the nails*
How you let go	*Slowly—quickly*
Even out pull of left and right elastic	*No alignment—careful alignment*

Beyond The Quick-Build

Activity 2, p. 120

Most Significant Variables	Suggested Modifications
Amount of pull on elastic	*Need repeatable scale*
Even lift and right pull of elastic	*Design for automatic connection*
Angle of launch	*Variable and accurate settings*
Release of projectile	*Design trigger release*

C

Activity 3, p. 127

Investigating Elasticity

Experimenting With Stretch

Band A: *about 3″ long; wide, stiff, short stretch*
Band B: *about 4″ long; narrow, long stretch*

Activity 3, pp. 129–132

Questions 5–10

5. a. Describe what you noticed about the points plotted for each rubber band.
   Same amount of stretch for each weight added

   b. Could a straight line include all or most of your data for each rubber band? Explain.
   Straight line fits most data, but not data near the end

6. Use your graph to predict the distance of stretch for masses smaller and larger than those you used.

   a. How accurate do you think these predictions will be? Explain.
   In the middle of the range, the data will be accurate; at the end, not so good.

   b. Pick a mass between that of two you used. For example, if you had used 45 g and 60 g in your calibration, choose a mass in between—perhaps 55 g. Figure out the length of the stretch corresponding to this applied mass.
   Band A: 0.2 kg is 0.4″; 0.3 kg is 0.6″; 0.25 kg will be 0.5″

   c. What do you think happens in the case of very large applied masses? Explain.
   Band breaks. Band is only so strong.

7. Try to express the relationship between mass and distance as a formula.
   For Band A, stretch = 2 × weight

APPENDIX C
SAMPLE ANSWERS

8. Hooke's law states that the stretch is proportional to the mass you applied. The region in which a material—in this case the rubber band—obeys Hooke's law is called the *elastic limit* of the material.
   a. What does your graph tell you about the elasticity of your rubber bands?
   Limit is near 0.5 kg; after that, stretch 2 × wt.

9. a. Plot both lines from the graphs you made in question 4 on the same set of coordinates.

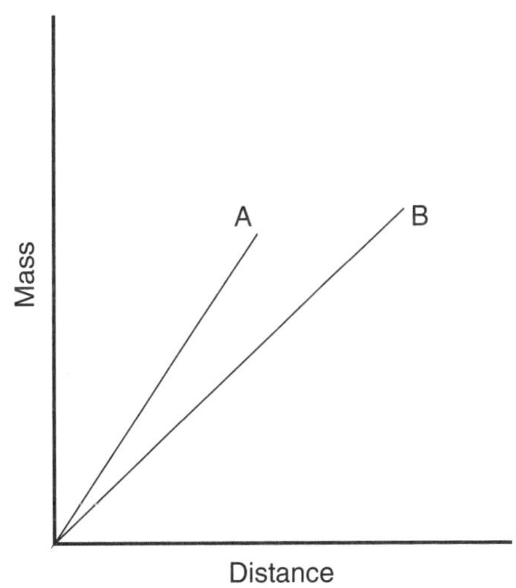

   b. Compare the two lines and explain similarities and differences between them. Be sure to compare them with regard to their conformity to expectations from Hooke's law.
   B stretches more than A
   c. What do you think the area under your plotted lines represents?
   Something to do with how hard you pull. (Note: Units are weight times distance. Force × distance = work; hence, units are units of energy.)

10. How could this exercise help you design your mechanical launching system?
    Can build scale to show how hard you have pulled on the rubber band.

C

Activity 4, p. 142

Beyond the Quick-Build: Second Pass

Most Significant Variables	Suggested Modifications
Lack of even left-right pull	*Center projectile on rubber band*
Projectile veers left or right	*Make groove on track*
Angle shifts at firing	*Build solid system for launch angle control*

Activity 4, p. 143

Team Feedback

Interfering Variables	Suggested Modifications
Slider is sloppy	*Redesign*
Pedestal not rigid	*Redesign*
Screws bend	*Tighten or design better attachment for rubber band*

Glossary

ballista: a stone-thrower catapult usually using a crossbow design

best-fit line: a smooth line drawn through the scatter of data points plotted on a graph leaving as many points below the line as above

brainstorming: a group problem-solving technique that involves the spontaneous contribution of ideas from all members of the group

calibration: determining a measurement scale or aligning a device with a measurement scale

catapult: a device used to throw things

dynamics: a branch of physics related to the effects of forces

elastic limit: the maximum stretch that can be applied before the stretch no longer obeys Hooke's law (usually damage is done to the elastic when it is stretched beyond this limit).

elasticity: the property that allows an object to stretch and return to its original length

energy transfer: the movement of energy from one place to another or from one form to another

forces: pushes and pulls

frequency distribution: a tabulation of the number of times events happen

iterative processes: processes that are repeated over and over; usually gradually approaching some goal

kinematics: a branch of physics related to motion

milestones: key markers along some process that provide a rough measure of progress toward some goal

onager: a type of catapult similar in design to that of the common mousetrap, in which a projectile is held in a rope sling to give it extra thrust

optimization: the process of discovering optimal conditions

optimize: to design for best results

pawl: a toothed wheel used in a ratchet

GLOSSARY

portfolio: a collection of products or designs

prototype: an original model; first full-scale and (usually) functional form of a new type of construction

ratchet: a device that allows an axle to rotate in one direction but not another; usually involves a saw-toothed wheel and a spring lever that catches the teeth so as to prevent backward rotation

siege: a tactic used in warfare where the enemy is surrounded and cut off from sources of supply and support

sinew: fibrous material that can be twisted or stretched

specifications: the precise details of an invention, plan, or proposal

spring constant: the constant in Hooke's law that relates the amount of stretch to the amount of force

tension: a force pulling or stretching a body

torsion: a force twisting a body

trebuchet, trebouchet: a catapult that works like an unbalanced seesaw

variable: an object or quality of changeable value

Suggested Readings

Chevedden, P. E., L. Eigenbrod, V. Foley, and W. Soedel. 1995. The trebuchet. *Scientific American* July: 66–71.

Corfis, I., and M. Wolfe. 1995. Artillery in late antiquity: Prelude to the Middle Ages. In *The medieval city under siege,* ed. P. E. Chevedden. Suffolk, England: Boydell and Brewer Press.

Gartrell, J. E. 1990. *Methods of motion: An introduction to mechanics*. Washington, DC: National Science Teachers Association.

Keys, C. W. 1994. The development of scientific reasoning skills in conjunction with collaborative writing assignments: An interpretive study of six ninth grade students. *Journal of Research in Science Teaching* 31 (9): 1003–1022.

Payne-Gallwey, R. 1995. *The book of the crossbow*. New York: Dover Publications.

Petroski, H. 1996. *Invention by design: How engineers get from thought to thing*. Cambridge, MA: Harvard University Press.

Richmond, G., and J. Striley. 1996. Making meaning in classroom: Social processes in small-group discourse and scientific knowledge building. *Journal of Research in Science Teaching* 33 (8): 839–858.

Soedel, W., and V. Foley. 1979. Ancient catapults. *Scientific American* 240: 150–174.

Warry, J. 1995. *Warfare in the classical world*. Norman: University of Oklahoma Press.

Part III

CONSTRUCT a GLOVE

Introduction

Integrating Science and Technology

Construct-a-Glove is aligned with the National Science Education Standards for process and content standards in both physical science and biology, as shown in the Standards and Benchmarks Connections table later in this section. Through inquiry and design, students develop conceptual understanding of heat energy transfer, cell metabolism, and thermal regulation. Because design activities motivate inquiry and inquiry informs design, students engage in the iterative processes of scientific inquiry and technological design through a variety of hands-on activities.

Schedule

The minimum time needed to complete the core unit is about 12 class sessions. More time will be needed if you choose to include the enrichment activities.

In Construct-a-Glove, students are given the Design Brief challenge and instructions for making a quick-build insulated glove (about three class sessions). During the research and development phases (a minimum of seven class sessions), students identify relevant factors and variables, design and conduct experiments, and contribute to a product development team. They develop a physical model of their design, test, measure, analyze their data, and redesign if necessary. Students search for combinations and configurations of materials that can improve the performance of their prototype glove. Teams critique each other's prototypes and learn to build on the experience and insights of other groups. In the Communication activity (two class sessions), the team composes a product prospectus.

Key Ideas

Each key science idea used in Construct-a-Glove is covered in a text reconstruction exercise in Appendix B (p. 265). The first exercise, on homeothermic processes, involves a very simple reconstruction and is intended as an introduction to text reconstruction.

Homeothermic Process (Maintaining a Constant Internal Body Temperature)

The measured and perceived warmth of a hand is related to its direct connection to the body's heat engine and the hand's relatively large ratio of surface area to volume. Variables that add complexity to an insulated glove system design relate to the multiple functions of the hand: an appendage for body cooling, environmental sensing, and manipulating objects.

Heat Energy Transfer Processes, Insulation Materials, and Dexterity

Student teams measure temperature change over time as a gloved hand is exposed to cold. Properties of various materials are explored for their effect on hand warmth and dexterity.

INTRODUCTION

Inquiry and Design

Experiments are designed by students to supplement "fair test" comparisons of several manufactured gloves. Students conduct "hands-in" research to determine combinations of glove materials that balance thermal effectiveness with dexterity for a specific function. Working in development teams to design and construct an insulated glove system prototype, students present their research and development effort in a product prospectus.

Student Portfolios

The following items can be accumulated in portfolios for summative assessment:

- Pretest: Snapshot of Understanding
- Initial questions: Design Brief
- Individual information search
- Brainstorming record: Identifying Factors and Variables
- Problem statements: Team Situation Analysis
- Research and results: Investigating Heat Transfer and Insulation; "Fair Test" Comparison
- Group process description: Inquiry Process and Design Process
- Development Assignment
- Team Feedback
- Evaluation of prototype: Reflections on Design
- Group summary documentation: Creating a Product Prospectus
- Posttest and self-assessment: Snapshot of Understanding

Part III

INTRODUCTION

Standards and Benchmarks Connections

Task	Source
Students build to specifications and observe and measure thermal performance of a simple insulated glove system. **Standard/Benchmark:** Materials and Manufacturing; Systems; Manipulation and Observation; Physical Science Content; Life Science Content; Unifying Concept: Evidence, Models, and Explanation	AAAS 9–12, NSES 9–12, NSES K–12
Students conduct independent information searches in context of contemporary human inventiveness. **Standard/Benchmark:** Issues in Technology; Nature and History of Technology Science, Technology, and Society	AAAS 9–12, ITEA 9–12, NCSS VIII
Students describe variation and identify variables and corresponding potential controls for improving design to meet performance goals. **Standard/Benchmark:** Systems; Manipulation and Observation; Technological Design; Unifying Concept: Evidence, Models, and Explanation	AAAS 9–12, ITEA 9–12, NSES K–12
Students use elements of inquiry, investigating heat transfer and insulation, to inform design. **Standard/Benchmark:** Scientific Inquiry; Design and Systems; Inquiry; Technological Design; Utilizing and Managing Technological Systems	AAAS 9–12, ITEA 9–12, NSES 9–12
Students use "fair test" comparisons to quantify relative handwear performance parameters. **Standard/Benchmark:** Critical-Response Skills; Physical Science Content Standard; Life Science Content Standard; Unifying Concept: Constancy, Change, and Measurement; Technological Design; Linkages	AAAS 9–12, ITEA 9–12, NSES 9–12 and K–12
Students conduct research on human homeothermic regulation and hand function as input to glove system design. **Standard/Benchmark:** Basic Functions; Life Science Content Standard; Unifying Concept: Form and Function	AAAS 9–12, NSES 9–12 and K–12
Students communicate orally and in writing their interpretation of this investigation and what variables to control in design development. **Standard/Benchmark:** Scientific Inquiry; Technological Design; Math as Communication; Statistics	AAAS 9–12, ITEA 9–12, NCTM II
Students apply abilities of iterative technological design, including brainstorming, research, ideation, choosing among alternative solutions, development, implementation, and evaluating consequences. **Standard/Benchmark:** Technological Concepts and Principles; Technological Design; Developing and Producing Technological Systems; Systems Operation and Feedback; Abilities of Technological Design; Unifying Concept: Systems, Order, and Organization	AAAS 9–12, ITEA 9–12, NSES 9–12 and K–12

INTRODUCTION

Task	Source
Students use tools and processes to construct and modify working models. **Standard/Benchmark:** Developing and Producing Technological Systems	ITEA 9–12
Students collect, represent, and statistically process test data to improve design prototype. **Standard/Benchmark:** Manipulation and Observation; Utilizing and Managing Technological Systems Statistics	AAAS 9–12, ITEA 9–12, NCTM X
Students create detailed product specifications and manufacturing instructions for their final design. **Standard/Benchmark:** Design and Systems; Manipulation and Observation; Math as Communication	AAAS 9–12, NCTM II
Students communicate quantitatively the technical construction specifications and performance characteristics for their prototype in a product prospectus. **Standard/Benchmark:** Communication Skills; Design and Systems; Technological Design; Math as Communication	AAAS 9–12, ITEA 9–12, NCTM II
Students articulate principles of biological and physical science used in glove system design. **Standard/Benchmark:** Critical-Response Skills; Science and Technology; Technological Design Linkages	AAAS 9–12, ITEA 9–12, NSES 9–12
Students self-assess their learning by comparing pre- and post-Snapshots of Understanding. **Standard/Benchmark:** Issues in Technology; Science, Technology, and Society	AAAS 9–12, NCSS VIII

Source Key:

AAAS = American Association for the Advancement of Science. 1993. *Project 2061: Benchmarks for science literacy.* New York: Oxford University Press.

ITEA = International Technology Education Association. 1996. *Technology for all Americans: A rationale and structure for the study of technology.* Reston, VA: ITEA.

NCSS = Task Force on Social Studies Teacher Education Standards. 1997. *National standards for social studies teachers.* Washington, DC: National Council for the Social Studies.

NCTM = National Council for Teachers of Mathematics. 1991. *Professional standards for teaching mathematics.* Reston, VA: NCTM.

NSES = National Research Council. 1996. *National science education standards.* Washington, DC: National Academies Press.

Part III

INTRODUCTION

Course Outline

Introduction

- Overview: Insulated Glove Design Brief
- Snapshot of Understanding

Quick-Build

- Overview: Quick-Build Insulated Glove
- Identifying Factors and Variables
- Team Situation Analysis: Reflections on Your Quick-Build

Research

- Overview: Research
    - » Team Assignment
    - » Individual Research Report
- Investigating Heat Transfer and Insulation
- "Fair Test" Comparison

Development

- Overview: Development
- Development Assignment
- Team Feedback
- Reflections on Design

Communication

- Overview: Communication
- Creating a Product Prospectus
- Snapshot of Understanding

DAY 1

ACTIVITY 1
Insulated Glove Design Brief

OVERVIEW: INSULATED GLOVE DESIGN BRIEF

In this activity you will be researching, designing, building, and improving an insulated glove system. You will use both technological design and scientific inquiry as processes to investigate and improve the performance of your prototype.

Design Challenge

As a member of a product research and development team, design an insulated glove system that keeps the tip of your index finger as warm as possible in uncomfortably cold surroundings, while maintaining dexterity for a specific function.

Scope of Work

- *Quick-Build:* Build and test an initial glove design according to instructions, and identify variables you can control to create an improved insulated glove.
- *Research:* Investigate heat transfer and insulation, and conduct a "fair test" comparison.
- *Development:* Specify function, redesign, build, and test; collect data and analyze patterns of results; then finalize your prototype for critical review.
- *Communication:* Present a product prospectus that summarizes your team's final design, including documentation such as sketches, data, specifications, and limitations.

Part III

ACTIVITY 1
INSULATED GLOVE DESIGN BRIEF

SNAPSHOT OF UNDERSTANDING

What I Already Know About Homeothermy, Heat Transfer, and Research and Development

The unit of study you are about to begin will challenge you to design, build, and performance test a prototype model of an insulated glove. To meet the performance specifications, you will have to investigate heat transfer physics, biological temperature regulation, and insulative effectiveness of materials and configurations. Before you begin, record a sample of what you already know by answering the questions below. This is not a test; rather it is a series of questions that ask about your current knowledge of key ideas in this unit. At the end of the unit, you will answer similar questions and compare what you have learned.

1. What are the parts of the hand?

   a. What are the functions of a hand (e.g., sensing, temperature regulation, manipulation)?

   b. Make a sketch of a hand and label the important parts and functions.

1

SciLinks — THE WORLD'S A CLICK AWAY

Topic: heat transfer
Go to: *www.scilinks.org*
Code: CAG01

Topic: radiation
Go to: *www.scilinks.org*
Code: CAG02

Topic: conduction/ convection
Go to: *www.scilinks.org*
Code: CAG03

2. List as many special-purpose kinds of gloves as you can. Place a "T" by those specifically designed to provide thermal protection. For example, a welding glove is designed to provide thermal protection so you would put a "T" next to "welding" in the list.

3. Think of a time when your hands were really cold. What were you trying to do?

How did you warm them?

Which heat transfer process did you use? (e.g., radiation, conduction, convection)

ACTIVITY 1
INSULATED GLOVE DESIGN BRIEF

4. What test(s) could you perform to determine if an animal is "warm-blooded" (homeothermic) or "cold-blooded" (poikilothermic)?

5. To maintain your relatively constant body temperature of 36°C, what does your body do automatically?

What are some things you do purposefully to make yourself warmer or cooler?

6. What are *temperature* and *heat,* and how are they related?

DAY 1

ACTIVITY 1 Teacher Pages

OVERVIEW: INSULATED GLOVE DESIGN BRIEF

Give students the Insulated Glove Design Brief and the Snapshot of Understanding. Initiate class discussion and highlight important design issues. In the Snapshot of Understanding, students write short answers to questions about their prior knowledge of heat transfer, body temperature regulation, and research and design processes.

Preparation

- Read and become familiar with the entire unit.
- Photocopy student activity sheets for distribution.
- Prepare an introduction to motivate student interest.
- Define your assessment system with a clear, simple description.

Materials

- Student Activity Sheets
    - » Course Outline
    - » Overview: Insulated Glove Design Brief
    - » Snapshot of Understanding
- Ring, Pocket, or Folio Binder (student supplied; for keeping student activity sheets, notes, and drawings for reference and portfolio assessment)

Time Requirement

This activity requires one class session. Completing the Snapshot of Understanding takes about 20 minutes.

Part III

ACTIVITY 1 TEACHER PAGES

Teaching Suggestions

Introduction

Hand out the Insulated Glove Design Brief student activity sheet. Ask students to keep these and future sheets together and to bring them to the classroom with other notes to serve as a record and reference for daily activity (and assessment) in the unit. Advise students that they will work in teams, using processes of technological design and scientific inquiry, and that other teams will critique their prototype with respect to the challenge criteria.

Students are also required to document their activity in order to contribute effectively to the final team presentation and enhance their individual portfolios. Be clear on your rubrics for assessing their work and share them with students. Indicate which activities will be individually graded and which will be given a team score. Be prepared to justify team scoring if some students (or parents) are not used to the idea.

Issuing the Construct-a-Glove Challenge

We encourage you to expand the challenge to accomplish additional learning objectives, but be careful to think through what will be involved. For example, a criterion of dexterity (such as the ability to pick up a marble with the insulated glove system) might be added to enrich the challenge if robotics or finger function is pertinent to your course objectives. But this addition will demand more sophisticated technical materials, involve greater construction difficulties, and require more student time.

Here are some other possible criteria for this challenge:

- Let students specify the dexterity they have achieved after the fact, in their product prospectus.
- Let teams choose to design for one of several simulated bid invitations. You will need to prepare the bid request document (be sure to include dexterity task specifications).
- Set one uniform standard for all teams to achieve with the gloved hand immediately following a standardized immersion time in ice water. Example

tasks include picking up a pencil off a flat surface, operating a camera, using pliers, holding a nail for hammering, tuning a radio, opening a keyed lock, and placing a nut on a bolt.

- Set a theme such as survival in snow country. If you use this theme or a similar outdoors survival theme, orient the challenge toward natural insulation materials such as leaves and grasses that can be gathered outdoors, and specify necessary hand functions such as gathering firewood and signaling for help. (Define or supply the range of options so as not to damage the local environment.)

Preassessment

Hand out the Snapshot of Understanding. Emphasize that it is not a test and that students will not be graded on this activity. The purpose of the Snapshot is self-diagnostic—to find out what students know initially about the key science and technology learning objectives of Construct-a-Glove.

An inventory of students' prior knowledge is an important teaching and learning tool. Not only does the inventory help guide students toward the concepts they need to learn the most, but it also prepares them to accept new information in a manner that ties meaningfully to what they already know. At the end of Construct-a-Glove, students will be able to compare answers to Snapshot questions given at the beginning of the unit with those they will answer at the end of the unit.

In the Snapshot of Understanding, students list as many special-purpose gloves as they can, given one example (welding). Here are some additional examples: food handling, meat cutting, dish washing, staining/painting/waxing, gardening, wood cutting, handling chemicals, working on electrical lines, firefighting, driving, traffic directing, fashion, surgery, cattle roping, archery, baseball, boxing, diving, golfing, hockey, hunting, ice fishing, mountain biking, mountain climbing, skiing, and spacewalking.

Encourage a class discussion of differences among special-purpose gloves; this can help students better relate form to function.

Part III

2 DAYS

ACTIVITY 2
Quick-Build Insulated Glove

OVERVIEW: QUICK-BUILD INSULATED GLOVE

Your body, like a home in cold climates, has a complex heating system optimized to maintain an internal temperature typically well above that of its surroundings. Chemical digestion of food can be compared to combustion of fuel in a furnace. Your circulatory system pipes heat to all parts of your body, just as air ducts or pipe loops do in the home. Just as some parts of your home feel warmer or cooler than others, temperatures at different locations in and on your body can vary within a range of comfort or concern. Your body's many biological sensors can reduce and redirect heat, and even activate evaporative cooling (sweating). Similarly, zoned thermostats control home heating and cooling by on/off switching or speed control of fans and dampers, or pumps and valves. Rooms, or hands, that are too cool can be warmed by either adding heat or reducing heat loss with insulation.

Work in teams of three to do the following experiments concerning thermal energy, heating, cooling, and insulation. Each individual should record independent notes and answers as well as the shared team data. These will be used later for portfolio assessment.

1. Decide which one of your team's available hands will be gloved and tested. Record in your notebook the factors you took into consideration when deciding which team member's hand you chose (e.g., volume-to-surface ratio, warmth to touch, rings, nail length) and the reasoning you used.
   a. Trace the chosen team member's hand on a sheet of paper.
   b. Mark points 1, 2, and 3 (as in the diagram on p. 208) and select two additional points 4 and 5.
2. Predict where you think the coolest and warmest temperatures will occur. Label these points C and W.
3. Measure and record the temperature of each of the five numbered points. Take or express your readings in degrees Celsius. Describe in your notebook any difficulties you had in measuring or uncertainties you have about your data.

4. Lightly tape the temperature sensor to the test hand at a location relevant to the design challenge (see, for example, the diagram below), and insert the test hand into one of the smaller and tightest-fitting disposable gloves provided. Immerse the gloved hand in ice water, being careful not to allow water inside the glove. Record temperature changes over a four-minute period.
   Summary of baseline findings:

5. Remove the gloved hand from the water and towel dry the outside of the glove. Apply a layer of insulation and/or spacer materials (your choice) to the outside of the gloved hand and carefully insert into a second, larger glove. Be sure to consider the issue of dexterity and the need to remove the inner glove intact.
6. Immerse the double-gloved hand in ice water, again being careful not to allow water inside the glove. Take new temperature readings over time and compare with your baseline data. Remove your quick-build glove carefully.

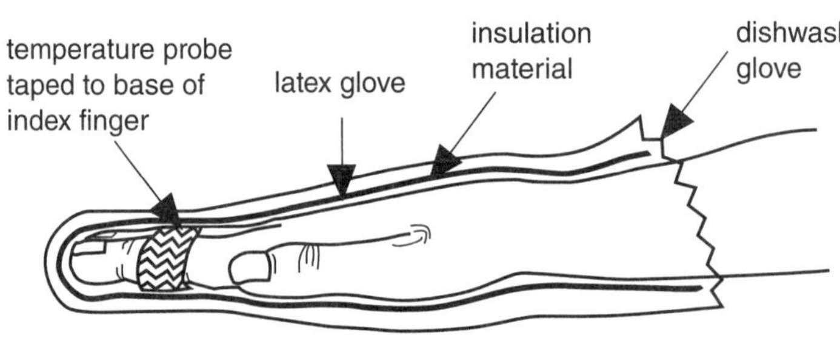

Part III

ACTIVITY 2
QUICK-BUILD INSULATED GLOVE

Identifying Factors and Variables

Reflect on the quick-build experience and identify the factors and variables that you think are important to meeting the design challenge. Classify the variables into two categories: those related to the glove system itself and those more associated with the user. Indicate the range of variation expected, quantitatively (numerically) or qualitatively (characteristically). Then list the corresponding design factors, constraints, or concerns you can anticipate. Be prepared to contribute your ideas as input to your class or team for the next research and development stage of your work.

Glove System Variables	Range of Variation
Example: *Convection seal at wrist*	← *no seal* *loose* *tight* →

User-Related Variables	Range of Variation
Example: *Starting state of user's hands*	← *chilled* *normal* *hot & sweaty* →

Design Factors	Constraints
Example: *Wrist seal pressure*	*Prevent convection without constricting blood circulation*

Team Situation Analysis: Reflections on Your Quick-Build

From a compilation of design factors and system variables, work as a team to select the key parameters you and other members agree to research. Assign each parameter to a team member who will be responsible for that aspect of the improved performance of your insulated glove. Record your selected variables, along with corresponding notes for modifying the quick-build, in the table below.

Most Significant Variables	Suggested Modifications
Example: *Direct heat loss through conduction*	*Slow heat flow with more layers of insulation*

Part III

2 DAYS

ACTIVITY 2 Teacher Pages

OVERVIEW: QUICK-BUILD INSULATED GLOVE

In a single, fast-paced class session students work in groups of three, following instructions to construct a quick-build insulated glove. They use an ice water bath to create a temperature gradient for rapid baseline and quick-build testing. They record data on hand temperature change over time, and make notes for improved experimental and product design. Homework and class discussion following the quick-build should stress identification of key variables.

Quick-Build

The activity is divided into six steps:

1. Selecting location of data points on the hand
2. Predicting relative outcomes of temperature readings at the selected points
3. Measuring the temperature of the index finger under normal, room-temperature conditions
4. Taking a baseline temperature measurement of the index finger immersed in an ice water bath
5. Designing and constructing the quick-build glove
6. Measuring the temperature of the index finger immersed in ice water but protected by the quick-build glove

Identifying Variables

As homework, students individually reflect on their quick-build experience and create a list of factors and variables to be considered. In the following class, these ideas are compiled into one list and then pared down to a key set via a class discussion. In support of this assignment, you may add context reading (as homework) where students read one or more illustrated industry handwear product descriptions from clothing or equipment catalogs (see p. 250 in Appendix A for an example). Potential contexts may span a full range of specialized student interests, from synthetic textile manufacturing to Arctic exploration to ski slope fashion. This will quickly broaden their awareness of design considerations and help them acquire a common vocabulary for discussions and further research.

Team Situation Analysis

Student product development teams analyze the class compilation, reflect on the factors and variables deemed most significant to meeting the challenge, and select agreed-upon areas to research.

Preparation

- Obtain quick-build construction and testing materials.
- Tailor quick-build instructions or assembly drawing to materials obtained. Reproduce, if desired.
- Preview library and internet resource information.
- Consider noise, safety, access to water, and cleanup in choice of location for the hands-on quick-build exercise.
- Designate and arrange assembly and testing areas for the quick-build.
- Organize materials for orderly access.
- Determine student team size (groups of three work well) and formation strategy.
- Contact related career representatives in science, manufacturing, history, social science, fashion, sports, and other fields for establishing context and relevance.

ACTIVITY 2
TEACHER PAGES

Materials

For Each Student:

- Student Activity Sheets
    - » Overview: Quick-Build Insulated Glove
    - » Identifying Factors and Variables
    - » Team Situation Analysis: Reflections on Your Quick-Build
    - » Homework: Individual Information Search (optional)
    - » Sample Industry Handwear Description (optional)
    - » Thermal Factoids (optional)

For Each Team:

- quick-build construction and testing materials, including scissors, staplers, adhesives, and tape
- digital thermometer, computer-linked temperature probe, or indoor/outdoor cable thermometer
- three to five disposable painter's, food service, cleaning, or surgical-type gloves in a range of sizes
- one cotton and one wool glove (or similar assortment) for layering
- scraps of fabrics and materials (cotton, wool, papers, foils, bubble wrap, foam sheet, yarn, straw, vermiculite, drinking straws, etc.) for extra insulation
- ice water
- containers
- sponges
- towels
- digital camera or video for recording team processes and design evolution

2

Safety Alert!

Be alert to safety during all tool and adhesive use sessions. If using latex gloves, provide the opportunity for any student to avoid exposure if they have a known latex allergy, and inform others of symptoms to watch for, including local swelling, hives, or rash. Severe reactions are rare and are almost always preceded by repeated minor reactions of gradually increased severity.

Students who have had previous cold injuries such as frostbite on their hands may have damaged capillary fields, which could both affect thermal reaction and cause return of discomfort to the student.

Time Requirement

This activity requires two class sessions: one to build and test the quick-build insulated glove and one to discuss variables and (if time allows) experiment with some trial-and-error modifications.

Teaching Suggestions

Day 1

Building, Testing, and Sketching the Quick-Build

This needs to be a fast-paced action class—a dramatic break from the reflective mood of the introduction. Allow students one class session to work in teams to build and test a quick-build insulated glove. Customize directions to the specific materials you are providing.

Arrange the common tools and supplies, such as scissors, staplers, adhesives, and tape, so that all students have easy access to them. Students may need reminders or a demonstration of safe and effective use. Intervention by you at critical junctures may also be required to ensure safety throughout the exercise.

Before the end of class, each student should make a rough sketch of his or her group's quick-build; for homework they should fully label each part. If appropriate for your class, you might recommend that manual drafting, computer-aided design (CAD), anatomical sketching, or another technical illustration process be used for later inclusion in the team report and individual portfolios.

ACTIVITY 2
TEACHER PAGES

Procedural Notes

- Students should start with the small gloves (and possibly the smallest hand), so that the larger glove(s) can be pulled on as outer layer(s) over any applied spacing and insulation materials.
- Each student team should resolve how to deal with rings and long fingernails and should present their rationale in answer to the question about hand selection on the Quick-Build Insulated Glove activity sheet.
- Taping the temperature probe to the pad of the index finger before hand insertion in the glove is recommended to ensure consistent skin contact; care must be taken not to use a tight wrap that constricts circulation, or an amount or type of tape that provides significant insulation.
- Most thermal gloves are not designed to be waterproof; use of the ice water bath is, however, the most convenient situation for rapid testing. A very thin disposable glove can be used as the outer waterproof layer to keep inner layers dry.
- A few students who maintain careful and continuous temperature change observations may see homeothermic feedback response (cold-induced vasodilatation). In this case, hand temperature rises after initial cooling.
- When a hand or foot is cooled to 15°C, blood vessels constrict and blood flow decreases. If cooling continues to 10°C, constriction is interrupted by periods of dilation with an increase in blood and heat flow. This "hunting reaction" recurs in 5- or 10-minute cycles to provide extremities minimal protection from the cold.
- If your course objectives or students' interests include bioregulation, you may wish to encourage further research on topics such as windchill, hypothermia, and other cold weather injuries such as frostbite, trench foot, and chilblains.

Homework: Identifying Factors and Variables

After the fast-paced quick-build class, homework provides a welcome time to reflect and determine next steps. Set the stage for design improvement analysis with the Identifying Factors and Variables activity sheet. On the sheet, students list key variables they observed as they built and tested their quick-builds, and they specify a range of possible variation. Stress the importance of completing this homework overnight so that it informs the next day's discussion. Students classify the variables into two categories: those that are part of the glove system itself and those more associated with the user. An example of each kind of variable is included on the student activity sheet. Some variables you might expect students to identify are listed below.

Glove System Variables

- the kind of covering material
- tightness of elastic or cuff seal
- the amount of trapped air
- amount and kind of insulation material
- finger fit
- seam integrity

User Variables

- user metabolism
- recent caloric intake
- hydration position
- contact of thermometer
- consistency of operator positioning for measurements
- user circulatory characteristics
- previous cold injuries
- user-installed insulation (body fat, hat on head, etc.)

Part III

ACTIVITY 2
TEACHER PAGES

- presence of caffeine (diuretic), alcohol (vasodilator), or nicotine (vasoconstrictor) in bloodstream

Students are also asked to list design factors that correspond with the glove system and user variables. Key design factors will vary with the specifics of the challenge but might include some from the following list:

- comfort
- static charge
- durability
- chemical resistance
- dexterity
- pressure on hand
- flexibility
- softness
- bulk
- stiffness
- warmth-to-weight ratio
- fit—form and size
- insulation loft
- venting
- cushioning protection
- cut resistance
- abrasion protection
- stretch
- permeability
- drying time
- color and visibility

Day 2

Class Discussion of Variables

Start a class discussion of the homework just completed. Compile a class list of the variables and ranges of variation students can identify. You may want to make a large chart on the board and jot down students' ideas so that all students have access to the complete list. Use the discussion to set the stage and level the playing field for the next activity where the student design teams plan modifications to improve their design.

Quick-Build Quick Changes

If time allows, students can experiment with some quick trial-and-error modifications to their quick-build to devise and conduct "fair test" hands-on experiments on the characteristics and configurations of insulation materials.

Stay alert for opportunities that may deviate from parts of the preplanned activities, especially those proposed by the students themselves. Pursuing students' promising alternative outcomes warrants careful consideration; not only do such detours accommodate student interest, they also serve as a valuable source of motivation.

ACTIVITY 2 TEACHER PAGES

Side Roads

Sample Industry Handwear Description

The Sample Industry Handwear Description in Appendix A (p. 250) can be used at any time to provide students with one example of the kinds of design considerations that real glove manufacturers use. Similar materials appear in various outdoor clothing catalogs.

Thermal Factoids

The Thermal Factoids sheet in Appendix A (p. 254) consists of a collection of short statements about thermal science that are designed to puzzle and provoke—and hopefully to stimulate interest. Section A has a series of short statements that should be identified as true or false; all are true. Section B provides an excellent take-home lab that can easily involve the entire family. It can also be conducted as an in-class demonstration, perhaps with a blindfolded student determining which container is hotter. Section C contains important scientific definitions and conventions that students may need to know to complete their designs.

Individual Information Search

One way students can learn more about design options for their insulated glove challenge is to research information about handwear on the internet (see Appendix A, p. 257).

ACTIVITY 3 Research

3 DAYS

OVERVIEW: RESEARCH

You have completed the quick-build insulated glove and observed some effects of limiting heat loss by decreasing conduction and convection. To improve your insulated glove system for practical use, you need to learn more about interactions between the thermodynamic and bioregulatory systems involved. You will also need to search out information on various materials and configurations.

There are different ways to approach and conduct research. You could use trial and error—simply try a lot of combinations of materials and see what works best or is acceptable. This could take a lot of time and require a good deal of materials gathering and data collection. A sounder investment of time, effort, and cost can be made if you conduct an initial information search.

For your initial research, you will probably use a library, the internet, textbooks, and personal queries. Then you will need to identify some questions and design experiments to answer them. Refer to the Inquiry Process resource sheet for more guidance. Realize that one question often leads to another, so you might need to repeat the inquiry cycle several times. You may even need to return for further research after you have otherwise moved on to developing and building a prototype physical model.

Team Assignment

Using the list you made of significant factors, divide up tasks among your team so that each factor will be fully researched. Make sure that each group member understands the key information needed. Working within the parameters given by your teacher, establish a schedule for individual reports that allows your team to review each report before your teacher asks to see it.

Individual Research Report

For homework, you will research the topics that your team assigned to you. You must present your team with a written report that details the most important things you discovered. But there is really no way to know now which parts of what you research will prove most useful when the team begins work on the prototype. For this reason, an important part of your research report is the identification of sources. A good report shows clearly where to look for the best information.

Part III

ACTIVITY 3 RESEARCH

Investigating Heat Transfer and Insulation

The research experiments you design for this activity should help you determine the best of available insulation materials and application systems to use in your insulated glove prototype. To arrive at a suitable experimental design, you might try listing anticipated problems in the procedure or interpretation, and then posing solutions. A couple of sample considerations are offered here to stimulate thought.

Problem: Metabolic and circulatory feedback effects vary among and within human hands, thus the temperature inside a glove depends on the hand as well as the glove.

Solution: Replace a radiating human hand with a water-filled glove at known initial temperature. Measure heat loss over time. How does this control of variables affect interpretation of your results?

Problem: Immersion in ice water provides fast experimental results but requires that outer materials be waterproof, thus unduly limiting material choices.

Solution: Adapt the experimental setup to use a source of constant cold air temperature, such as a refrigerator or an air conditioner. Or use a thin, waterproof, disposable film glove as an outer layer and devise a correction factor for the insulating effect.

SCILINKS® THE WORLD'S A CLICK AWAY

Topic: metabolism
Go to: *www.scilinks.org*
Code: CAG04

Topic: insulation
Go to: *www.scilinks.org*
Code: CAG05

1. Write down your research protocol for evaluating gloves.

2. Sketch your experimental setup and label it so that another researcher would be able to duplicate your experiment(s).

"Fair Test" Comparison

A "fair test" comparison is a comparison among two or more designs in which no design is given an advantage. Follow this procedure to make product comparisons:

1. Observe differences between your quick-build insulated glove and a commercial glove that has been designed to be worn in cold weather.

Quick-Build Glove	Commercial Glove

2. Using your research protocol, compare the insulation effectiveness of the commercial glove with that of your quick-build product.
3. Do the results of your research protocol fit with your understanding of how the two gloves compare?
4. Decide whether changes need to be made in your protocol. Is it a fair test?

Part III

3 DAYS

ACTIVITY 3 Teacher Pages

OVERVIEW: RESEARCH

Students work in their product research and development teams to investigate significant variables in heat transfer and insulation. They divide the tasks of researching and preparing individual research reports.

Students use controlled exposure tests (with an ice water bath) for experiments they design. To control for homeothermic feedback and other user variables, they can use a water-filled glove instead of a human hand, measuring temperature change of the water in the glove with different insulation, layering, and spacing configurations. Results should be presented in a data table and a graph to compare the amount of temperature change as a function of time for each of the different test configurations.

Each design team creates a research protocol for testing gloves. The teams validate this protocol with "hands-in," "fair test" comparisons of commercial insulated gloves. They consider their lab findings in the contexts of human thermal regulation, materials science, insulation technology, and handwear design.

This is a good time for an optional teacher presentation on heat capacity and specific heat, as well as composite insulation factors for different combinations of materials (see For the Class section of Materials on pp. 224–225).

Preparation

- Devise ways of fixing gloves to hang vertically for the water-filled experimental setups.
- Predetermine suitable type, quantity, and variety of test materials for safe and manageable team use of class time.
- Consider allergies, safety, cost, time, mess, and cleanup.
- Locate and obtain materials in quantities sufficient for both the research and development activities.
- Encourage students to provide additional materials, but prepare guidelines and implement a process or practice of approval before use.

3

Materials

> **Safety Alert!**
> Most adhesive materials present some hazard from fumes or skin contact. Avoid the worst—such as superglues and spray adhesives—entirely.
> Read labels and have specified solvents and cleanup materials available.
> Instruct and caution students regarding proper and careful use.

For Each Student:

- Student Activity Sheets
  - Overview: Research
  - Investigating Heat Transfer and Insulation
  - "Fair Test" Comparison
  - Inquiry Process (from Appendix A; optional)
  - Homeothermic Regulation (from Appendix A; optional)

For Each Team:

- clamps and stands (or duct tape) for stabilizing each glove in a vertical position for testing and measuring (if using water-filled gloves to control for homeothermic feedback effects)
- thermometers (digital indoor/outdoor with waterproof flex cables)
- ice, water, and buckets (3 lb. coffee cans serve well)
- timing devices
- towels
- graph paper
- assortment of commercial insulated gloves provided by students

For the Class:

- A diverse selection of potential insulating, reflective, spacer, and encasing materials:
  - a range of sizes and types of disposable gloves
  - Styrofoam peanuts
  - newspaper

ACTIVITY 3 TEACHER PAGES

- soilless potting mix
- beanbag chair fill
- cotton bat
- bubble wrap
- aluminum foil
- plastic wrap
- drinking straws
- straw, excelsior
- Easter "grass"
- coin tubes
- clothes drier lint
- Cheerios

Students may also be encouraged to add to the collection with various adhesives such as double-sided cellophane tape, glue sticks, rubber cement, duct tape, foam adhesive pads, and silicon sealant.

Time Requirement

This activity requires three class sessions. In the first session, students investigate potential insulators and design experiments for measuring glove insulation. In the second session, students discuss the graphs they have created in connection with the experimental setup and glove configuration. In the third session, students compare insulative qualities of commercial gloves and discuss the effectiveness of their designs.

Teaching Suggestions

Day 1

Group students into product research and development teams of three to five. Review their assignment: Investigate different insulators and design a uniform method for measuring the thermal insulation of gloves.

Each team of students will likely test different factors and reach different conclusions, but if given the opportunity to share their experimental designs and findings, all will benefit by the diversity of multiple team approaches. Comparing findings across the class should reinforce appreciation of the numerous variables (and possible solutions) involved. You may choose to eliminate the need for each team to test all key variables by assigning certain key variables to specific teams.

Using the quick-build glove as a starting point for the experimental setup makes the data collected earlier directly relevant as a baseline for assessing improvement through redesign. If possible, students should complete their data collection and organize their results in a table during this class session. Use your judgment to decide whether to extend, curtail, or defer research activity beyond this time allotment. Additional time for research and experimentation may fit in better during the Development activity (Activity 4).

Outside of class, students should construct a graph for each set of data in their table and answer the series of homework questions on the activity sheet.

Day 2

For the first part of the session, facilitate the previous homework review by choosing several groups to sketch or project their graphs on the board for use as a basis of discussion. Ask students to detail the experimental setup and glove configuration associated with each graph. To assess the level of class understanding, pose questions and elicit responses from a range of students. Sample questions might include:

- Which graphs tell us about layering as an insulation factor?

ACTIVITY 3 TEACHER PAGES

- Why are the graphs different? (relate to observed properties of each configuration or material, such as amount, thickness, thermal conductivity, and entrapped air)
- How are the relative thermal properties of different materials shown in these graphs?

For the remainder of the session, guide the class in choosing a common research protocol so that the work of different teams can be fairly compared.

Day 3

In this session students integrate the idea of feedback into the way they have been thinking about homeothermic regulation. Include a discussion about heat flow as the cause of the sensation of coldness. If you touch a metal object and a wooden object both at room temperature, the more conductive metal object feels colder because heat flows away from your finger faster.

The main activity for this session is to allow students to compare insulative qualities of various commercial gloves and describe the design effectiveness of each.

Because most gloves are not waterproof, students may use a disposable glove as an outer sheathing to enable the use of the ice water bath. They should test with the same team member's hand inserted in each glove and record thermal and other sensory data to compare with the earlier quick-build data set.

Assessment

Assessment may be based on whether students state testable hypotheses, follow through and document a controlled variable experiment, graph and write interpretations of results, record reflections on areas for improvement, and contribute to their team effectively.

Side Roads

Inquiry Process

The Inquiry Process sheet in Appendix A (p. 258) provides one possible way to think of the inquiry process and may be used either as an example or as the template for all your students to follow. If you use it as a template be sure to indicate that this is your choice and not the only possible way to view inquiry. Point out to students that there are many situations where it is useful to have everyone follow a common system even though many other systems would be equally good.

Homeothermic Regulation

The Homeothermic Regulation reading in Appendix A (p. 260) provides a more detailed account of the same process described in the Introduction (Key Ideas section). You will find this reading particularly useful if principles of biological science are important course objectives.

Part III

DAYS 4

ACTIVITY 4 Development

OVERVIEW: DEVELOPMENT

You have completed activities investigating insulation and conservation of heat with various materials in different configurations. You have also researched and observed the human thermal bioregulatory system response to cold. To meet the Construct-a-Glove design challenge, you will next develop an insulated glove prototype that minimizes heat loss while maintaining dexterity for a specific function.

Scope of Work

- Redesign your glove system for improved performance within specifications.
- Build your prototype and test its performance using your research protocol.
- Measure and record prototype test conditions.
- Compare your prototype with that of other teams.

Good planning is essential to good design. Look again at your research where you identified variables and key directions for further investigation. Think about how best to combine physical design options with biological considerations. Be creative within your objectives and constraints. Review the Inquiry Process (from the Research activity) and Design Process sheets for ideas on next steps. Evaluate your prototype critically and make modifications until you are satisfied with improved performance or run out of time. Remember to reflect on your process and record your reflections in your notebook; how you go about your design and what you learn are key elements in the self-assessment and communication activities to follow.

Development Assignment

Design Prototype

- Make a sketch or blueprint (use standard format)
- List all materials
- Write assembly instructions

Fabricate Prototype

- Collect or construct parts
- Assemble prototype

Test Prototype

- Apply research protocol

Evaluate Prototype

- Collect user reviews and other "informal" evaluation (see Team Feedback and Reflections on Design activity sheets).
- Using all available information, list strengths and weaknesses of your prototype, and compile a list of possible improvements.

Return to "Design Prototype" Step if There Is Time.

Part III

ACTIVITY 4
DEVELOPMENT

Team Feedback

Identify those factors that are interfering with the ideal performance of the prototype glove. Make suggestions about modifications that could improve performance. Record the problem factors and suggested modifications in the table below.

Factors	Suggested Modifications

Reflections on Design

1. What is the one feature of your insulated glove system about which you are most proud?

2. Describe the concerns you have with the present design.

3. Describe in a paragraph or two the approach your team took in the designing and building of your prototype. Include a problem your team encountered with the performance tests and how you went about solving it.

Part III

DAYS 4

ACTIVITY 4 Teacher Pages

OVERVIEW: DEVELOPMENT

As homework, students can read a brief description of the design process (on p. 263 in Appendix A; optional) to determine where in the activity loop they've been, where they are now, and where to go next.

On the quick-build glove diagram provided, students individually superimpose, label, and note additional planned performance enhancement features (e.g., wrist air seal, heat-reflective lining, solar-absorptive covering) determined from their research, then compare with team members to move toward consensus on selection of design elements.

Students construct a prototype of their improved design and document with a drawing in a format appropriate for the illustration technology used.

Students do a second pass on identifying variables that impact performance, this time for their redesigned glove, and record their suggested modifications. They consider changes required for mass production and record steps for replication efficiency.

Each team pairs with another for feedback. They compare glove features and make and receive suggestions for modifications.

Students performance test their prototype using a classwide common test protocol (agreed to ahead of time). They measure temperature change, describe patterns, and record all other relevant factors.

Students fill out the Reflections on Design activity sheet, describing strengths and concerns as well as their approach and team process.

Preparation

The Fabrication Zone

Consider quantities of work space, materials, tools, fasteners, and available adhesives, and optimize spatial arrangements for safe, efficient access.

The Performance Testing Arena

- If space in your room allows, give each group its own separate area to performance test its prototype. If there is not enough space for this, designate a "testing station" (near a sink or drain if possible) for groups to share in rotation.

- You may want students to prepare the test station configuration and to determine use and cleanup guidelines by class consensus.
- Consider which technological tools you might add (e.g., CAD). If your students are already proficient with computerized temperature probes, spreadsheets, or graphing calculators, the data they collect during performance testing can be entered, manipulated, and printed.

Materials

For Each Student:

- Student Activity Sheets
  - » Overview: Development
  - » Development Assignment
  - » Team Feedback
  - » Reflections on Design
  - » Identifying Factors and Variables (from Activity 2; for review)
  - » Inquiry Process (from Appendix A; for review if used in Activity 3)
  - » Sample Industry Handwear Description (from Appendix A; optional)
  - » Design Process (from Appendix A; optional)
  - » Team Situation Analysis: Reflections on Your Quick-Build (from Activity 2; optional)

For Each Team:

- clamps and stands
- disposable vinyl, latex, or poly gloves
- safety glasses or goggles

ACTIVITY 4
TEACHER PAGES

- one thermometer
- ice
- water
- bucket

For the Class:

- scissors
- staplers
- weighing apparatus
- hot-melt glue
- adhesive tape
- liquid latex
- insulating materials
- reflective materials
- layer separation materials (yarn, straw, etc.)
- cover materials
- graph paper
- timing device
- towels
- cornstarch or powder

Time Requirement

This activity requires four class sessions. Students should be encouraged throughout this sequence to continue their team collaboration and to do additional work outside of class. Indicate that completing the task on schedule requires considerable discipline.

Teaching Suggestions

Design is not linear: team processing times will vary. Nonetheless, development activities can be completed in four efficiently used class sessions—but only if you keep the pressure on.

Discussion of Homework: Design Process (optional)

Discuss the homework in the Design Process activity sheet. A good initial question to pose is, *Does the diagram depicting the design process reflect the experience in designing and building you have been having?* Other questions to ask are ones from their homework *Which elements of this process have you already experienced?* and *Where in the process do you think you are now?*

You may wish to make the criteria you will use to assess students' technological design capability available at this point and invite discussion.

Suggest to students that if they become stuck at any point, they should refer again to their portfolio sheets for ideas about what to do next. Highlight or elicit examples of how students already used or might soon use the design process loop in nonlinear progression. Students may wish to record scenes illustrating their group process with photographs or video.

The Design Process sheet provides one possible way to think of the design process and may be used either as an example or as the template you want all your students to follow. If you use it as a template, be sure to indicate that this is your choice and not the only possible way to view design. Point out to students that there are many situations where it is useful to have everyone follow a common system even though many other systems would be equally good.

Starting to Redesign the Improved Glove

Students may choose to proceed with the redesign of their gloves either by improving on their quick-builds or by starting fresh. Depending on their choice, students should (a) preserve or build on modification conclusions reached when working with the quick-build system in the research phase or (b) consciously begin anew on a distinctly different system, in which case initial research must be inserted in the development

Part III

ACTIVITY 4
TEACHER PAGES

process. Advise them also to keep the selected or specified purpose of their glove constantly in mind to guide decisions.

Resources for Redesigning

Before students begin redesigning, they might individually review and then discuss as a team the Sample Industry Handwear Description (in Appendix A) to define the team's design vocabulary. They could also review the Team Situation Analysis student activity sheet (from Activity 2) to remind themselves of modifications they identified prior to their research.

Timely Feedback

As students move beyond their quick-builds, it is essential that teams be allowed to grapple with the challenges of designing, building, and testing, and revisiting these processes. It is important, however, that students receive input at critical times.

One source of input is the team itself. Encourage brainstorming as an official design process to be performed by the rules. The Team Feedback activity gives every team a chance to share and explore ideas beyond those of the members.

Another important source of feedback is timely intervention from the teacher. Very often, a suggestion about how to solve a problem, or a reference to a diagram or picture, will stimulate action. Drawings or some sample student-built prototypes from previous classes may be helpful to stalled teams. But judgment must be exercised to avoid intervening too early.

Evaluation and Reflection

Just as students previously paused to evaluate and reflect on their quick-builds, they do the same again with the Team Feedback activity sheet, after they have had a chance to struggle with the redesign of their gloves or are confused and cannot continue.

You may wish to use the Team Feedback sheet to encourage one team to assess the work of another. For this reflection/assessment, pair each team with another for the purpose of providing feedback about each other's designs. Suggest to partner teams that they demonstrate their prototype to each other and that each team collaborate to provide written feedback to the other on the Team Feedback activity sheet. The goal here is twofold: for students to share ideas and for students to obtain helpful feedback from others.

Once groups have consulted and provided feedback to each other, provide some time and space for each team to discuss and decide their next steps. At this stage, you can assess their troubleshooting or problem-solving ability to conceive of modifications that address the problems they have been encountering.

Collecting Data

To test their prototypes, students immerse their gloves in ice water and record inside temperature change over time using a common "fair test" research protocol.

Deciding How Good Is "Good Enough"

Once groups start testing the performance of their prototypes, it is important to discuss with them how to know when to end the improvement cycle. The answer to this question depends on your teaching goals, any criteria agreed to by the class, and what is realistic given the materials with which students are working. In addition, the depth of development should be based on sustaining interest in view of their prior experience and skills as problem solvers, designers, and builders.

Whatever the standards you set for your students (or they set for themselves), remind them of the need to analyze and improve both the glove system design and the consistency of performance testing techniques in order to obtain more reliable results. As your students are testing their glove systems, look for evidence that they are evaluating and using their findings to make both product system and measurement process improvement decisions.

Design Self-Assessment

As a final development activity, students reflect on what they have done thus far in the Reflections on Design activity sheet. They are asked to identify positives and negatives about their current design and then to briefly describe their team approach and process. Because this activity sheet is brief, it gives you the opportunity to ask additional

ACTIVITY 4 TEACHER PAGES

questions. Science, math, and technology students might list or describe anticipated problems of actually producing their devices in mass quantities and various sizes to match their projections of consumer demand.

Assessing Students' Design and Build Capabilities

There are several key elements of students' design and build capabilities you could assess throughout the development phase:

- *How well are students able to develop solutions?* Developing ideas through to workable solutions is at the core of technological design. Look for evidence of students' ability to do this, both in your ongoing observations of students at work and in the responses they give to the five steps in their Development Assignment activity sheets.
- *How well are students able to evaluate the processes they have used?* This includes the extent to which they are able to identify strengths and weaknesses of their prototypes. Look for evidence of students' ability to do this, both in your ongoing observations of students at work and in the responses they give on the Reflections on Design activity sheet.
- *To what extent do students exhibit ownership of the task? Did it change with time? How much initiative did they take?* Look for evidence of this in your ongoing observations of your students.

DAYS 2

ACTIVITY 5
Communication

OVERVIEW: COMMUNICATION

You have completed the research and development (R&D) phases of the Construct-a-Glove challenge. You have redesigned, built, and tested an insulated glove system. But you will not be recognized for your effort without effectively documenting and communicating what you have done.

One measure of success of your R&D team is whether you have applied the criteria and specifications to come up with a useful, marketable product. In the marketplace it is the comparison with other products that determines success. Sometimes this comparison is based on measurable differences; other times, it is based on an image that has been effectively communicated to the buying public. To decide how to market your glove, realistically assess the strengths and limitations of your product and compare it with that of other teams. Based on these factors, determine a practical market niche for your product.

Scope of Work

- Present your redesigned insulated glove system in a product prospectus which outlines the rationale, substance, and outcome of your effort.
- Specify the materials and performance parameters for your prototype product.
- Write a suggested marketing plan for launching your product, or select a "next steps" topic to write about as an optional activity.

Part III

ACTIVITY 5
COMMUNICATION

Creating a Product Prospectus

The final activity is for you to communicate the results of your work in a product prospectus. Your goal is to include information that would interest and inform others about your glove system.

An important skill for you to demonstrate in your product prospectus is the ability to communicate clearly about your glove system in writing. The writing of each section should be well organized and clear enough for someone unfamiliar with your team's product prototype to understand.

You may find it interesting to consult several real product brochures before you begin, but don't feel that yours needs to be just like these. Be original, thinking about better ways of describing your team's glove as a system.

Product Prospectus Table of Contents

Consult the table of contents shown below for a list of the topics you should address in your product prospectus. Some parts of the prospectus, like your sketches and specifications, have already been done. Other parts, like the discussion of scientific principles, still need to be written.

Table of Contents

I. Guiding Principles

II. System Overview
   A. Parts and Materials Specifications
   B. Performance Specifications

III. The Science Behind Our Product

IV. Appendixes (optional)
   A. Product Comparisons
   B. Care and Cleaning
   C. Disclaimers and Warranty

Members of your team may wish to divide the responsibility for each of the sections that need to be written. Before you do so, read the suggestions below and clarify what each section will be about.

Suggestions for Product Prospectus Sections

Guiding Principles: In this section you should state any specialized activities or functions that your insulated glove system supports. Take into consideration what level of movement and dexterity your design provides and what activities such handwear allows. This section should be a total team effort so that sections that follow are unified by reference to the agreed purpose. First brainstorm as a team, then assign the actual writing—based on the brainstorming notes—to one member.

- **System Overview:** This section can include a sketch of your glove system with labeled parts.
    - » The *Parts and Materials Specifications* section could list details such as size, shape, composition, and quality of each component part.
    - » The *Performance Specifications* section is an appropriate place to describe the capabilities of your product. What are the minimum and maximum temperatures, wind conditions, levels of compression, and other conditions that your glove can handle and still provide hand comfort? How confident are you of your projections? (Your actual range of performance tests may have been restricted by time and opportunity.)
- **The Science Behind Our Product:** In this section you can discuss the scientific principles involved in the materials and configuration of your handwear. Consult and reference any appropriate sources to clarify your points persuasively to potential consumers.
- **Appendixes Are Optional:** Use them if you wish to include any additional information such as the following:
    - » For *Product Comparisons*, describe each team's product in terms of its strengths and weaknesses. Think of yourself as an employee of a large

ACTIVITY 5
COMMUNICATION

handwear industry manufacturer and your team's glove system as one in a range of choices for users with differing needs.

» In the *Care and Cleaning* appendix, you might let the user know what kind of periodic treatment you think your glove product will need to continue performing well.

» The *Disclaimers and Warranty* appendix addresses what your team as manufacturer will be responsible for should some failure occur with regard to durability, performance, and other factors in comparison with assurances and claims.

SNAPSHOT OF UNDERSTANDING

What I Now Know About Homeothermy, Heat Transfer, and Research and Development

You have met the challenge to design, build, and performance test a prototype model of an insulated glove system. To meet the performance specifications, you investigated heat transfer physics, biological temperature regulation, principles of insulation technology, and design process. One of your first steps was to record a snapshot of your understanding about key topics in this unit. By answering similar questions below, you can compare what you know now with your previous answers and self-assess what you have learned.

1. What are the parts of the hand significant to the design of an insulated glove with specific dexterity? Make a sketch and label its parts or describe in words.

2. List as many special-purpose kinds of gloves as you can. Place a "T" by those specifically designed to provide thermal protection. For example, a welding glove is designed to provide thermal protection so you would put a "T" next to "welding" in the list.

3. Describe what process your team used to design your insulated glove.

ACTIVITY 5
COMMUNICATION

4. Describe several key differences between *homeothermic* and *poikilothermic* animals.

5. Describe mechanisms and behaviors for human body temperature regulation.

6. Distinguish between *temperature* and *heat* and describe how they are related.

Next Steps

Choose one or more of the following extension topics and commit your thoughts in writing.

1. Present your views on the historical significance and impacts of gloves on human accomplishment.
2. Review the diversity of specialized gloves in the context of contemporary human inventiveness.
3. Describe promising market niches for new specialized gloves that might be developed to improve human capability, safety, comfort, and other factors.
4. Outline a comprehensive marketing plan for launching your team's glove design into a profitable product cycle.

Part III

DAYS 2

ACTIVITY 5
Teacher Pages

OVERVIEW: COMMUNICATION

Student teams summarize their learning by creating a product prospectus that includes information about specific applications for their glove system, its construction, principles of operation, and performance parameters. The Creating a Product Prospectus activity sheet provides a set of topics for students to address, along with explanatory text. In the final assessment, students answer questions similar to those at the beginning of the unit.

Preparation

- Consider spectator invitations if opting for oral presentations or for market introduction of prototypes.
- Provide (or have students bring in) a few different examples of product marketing media.
- Customize the product prospectus table of contents to fit the time and education objectives for your class (or encourage students to make such decisions).
- Arrange for use of word processing/graphics (or CAD) computer stations.
- Prepare a grading plan for your evaluation of the team and individual effort.

Materials

For Each Student:

- Student Activity Sheets
  - » Overview: Communication
  - » Creating a Product Prospectus
  - » Snapshot of Understanding
  - » Next Steps (optional)

For Each Team (optional):

- flexible seating to facilitate team discussions and writing
- word processing/graphics stations and presentation tools

5

Time Requirement

Two class sessions are required for this activity: one class for team writing of the product prospectus writing, with additional time needed as homework; and one class for team product prototype presentation, reflection, and final self-assessment. Allow about 20 minutes for students to complete the final Snapshot of Understanding, plus a brief time for them to compare their answers with those on their pre-unit Snapshot.

Teaching Suggestions

Discuss with your students which topics to include in their product prospectus and suggest that they develop a plan for delegating tasks within their teams to get the work done. Each team begins work with group brainstorming of possible intended uses or markets for their glove system. Suggest to teams that they take into consideration what their design's strengths and weaknesses are compared with other team's products. Encourage team members to review each other's contributions before assembling their final document.

An important skill for students to demonstrate in their product prospectus is the ability to communicate clearly in writing and graphics. Remind students that in both science and technology, communication is key to success. Your grading rubric should emphasize content, organization, and clarity.

Next Steps

In many ways, Construct-a-Glove is just a beginning in the process of examining gloves. If there were enough time, a great deal more could be done. The Next Steps section of the student pages suggests a few projects students could undertake after the unit. You will probably be adding your own options to this list each time you teach the unit.

Part III

APPENDIX A
Side Roads

The material in this appendix is intended to support activities that you may choose to add to those described in the core unit. Many of these are key activities, but they have been placed in this appendix because they can fit in several different places in Construct-a-Glove—exactly where they are used is a matter best decided by you in response to student questions and feedback. Some activities may be profitably used more than once. An analysis of the design process, for example, will provide different insights when used in the research activities than when used in the development activities.

In this appendix:

- Sample Industry Handwear Description
- Thermal Factoids
- Homework: Individual Information Search
- Inquiry Process
- Homeothermic Regulation
- Design Process

A

Sample Industry Handwear Description

Read the following glove description* and independently create your own one-page flyer highlighting features of your team's quick-build glove. Use drafting, sketching or illustration methods, and labeling styles as assigned by your teacher.

*Reprinted with permission of Outdoor Research, Inc., Seattle, WA, as posted online at *www.orgear.com*, last visited March 2000.

Expedition Modular Gloves: Technical Information

Expedition Modular Gloves can be used in a wide range of mountaineering, skiing, and cold weather adventure situations. It is an extremely warm glove with excellent dexterity for those super cold ice climbing days in Canada or lift skiing in deep powder in the Rockies. Any time dexterity is important, but you can't afford to lose much warmth in the process, this is the handwear of choice.

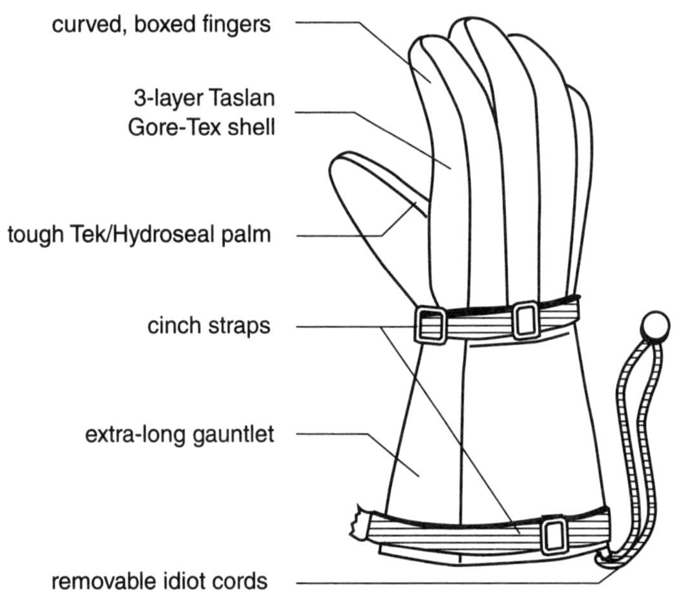

Part III

APPENDIX A
SIDE ROADS

Technical Features and Benefits: Shell Specifications

Fingers and Thumb Strongly Curved

The curved shape of the fingers and the thumb follows the position of function of the hand. This reduces the work the hand must do to hold an ice tool, ski pole, or ascender. It also lets the hand relax in its natural position without resistance.

Box Construction

The boxlike construction gives each finger its high volume, resulting in lower pressure on the hand and therefore a much warmer glove by design.

Three-Layer Taslan Gore-Tex With Laminated ToughTek/Hydroseal Palm

Gore-Tex is used to achieve the best possible waterproof, vapor-permeable, and windproof combination. Due to the complex construction of these gloves, the seams are not factory sealed, so the shells are not considered waterproof, but with the addition of the GT Liners (see description below) the hand can be kept fully dry. For most mountaineering and skiing situations it is as much durability as a person will need. Highly windproof and vapor-permeable.

Cinch Strap Adjustment at Wrist and Cuff

Velcro at the wrist keeps the glove from slopping around or sliding off the hand. It is placed slightly forward of the narrow point of the wrist. This is a deliberate design feature. It is more effective at keeping the fingers in the ends of the mitt. The cinch strap at the cuff keeps snow and wind out of the mitt and gives the wearer the option of venting excess vapor when working hard. Quick one-handed operation and no flapping strings. When closed, Velcro Cinch Straps do not flap around in the wind like typical draw cords. This can be important when working with the hands near the face as in ice climbing. Also nice when skiing.

Form-Fitting Thumb

In the coldest conditions it is imperative that a glove does not put pressure on the hand at any point, especially at the thumb. For this reason the Expedition Glove is constructed out of several separate pieces of material to create a high-volume, form-fitted section. This is one of the keys to the great warmth of this glove.

Extra Long Gauntlet

The gauntlet is long enough and big enough to accommodate even the poofiest down parka sleeves. For the nastiest blowing spindrift this is the only way to go. Combined with the cinch straps, the gauntlet creates a seal that is impenetrable, even when you're skiing with the aid of a snorkel.

Removable Idiot Cords

When the mitts are removed for manipulation of small objects, these cords act as keepers, allowing the wearer to dangle the mitts from the wrists. Once you get dialed into this feature you will quickly want to add it to all of your handwear!

Technical Features and Benefits: Liner Specifications

Standard Moonlite Pile Liner

- High warmth-to-weight ratio.
- Extremely durable, maintains its loft well.
- Fast drying.
- Palm and finger side walls are one layer of pile for dexterity.
- Back of hand and back of thumb have two layers of pile for extra warmth.

Poron Foam for Back of Hands at Knuckles

Protects knuckles from being bashed while ice climbing or skiing.

Flat Seams at Palm

Flat seams allow excellent "touch" and dexterity of the fingers.

Seam Allowance on Outside of Liner

The extra material from the seams is placed on the outside, away from the hand, for warmth and comfort.

APPENDIX A
SIDE ROADS

Strongly Curved Box Construction

- Follows the position of hand in the resting position.
- Matches the curve of the shell.
- GT Liner
- The GT liner is a single liner of Moonlite Pile covered by Gore-Tex and a final layer of WickLine. It lacks the Poron Knuckle guard of the standard but retains the curved shape and box seam construction.
- Totally waterproof for use in sloppy conditions.

A

Thermal Factoids

A common misconception is to confuse heat and temperature or to believe they are the same. Test your own understanding by marking "T" for true or "F" for false next to each of the statements in Section A below. Section B provides a quick experiment you can do yourself, or alternatively, you can make an outcome prediction and verify it with another classmate. Section C contains true information for reading and use in your research and development.

Section A: Temperature Versus Heat

____ When a material is hot, it has more thermal energy than when it is cold. Thermal energy is the total potential and kinetic energy associated with the random motion and arrangement of the particles of a material.

____ Temperature is the hotness or coldness of a material.

____ Heat is thermal energy that is absorbed, given up, or transferred from one body to another.

____ The temperature of a body is a measure of its ability to give up heat to or absorb heat from another body. Heat will flow from a body with a higher temperature to a body with a lower temperature, even if the cooler body contains more thermal energy.

____ Temperature is a physical property that determines the direction in which heat energy will flow between substances.

____ There is no instrument that directly measures the amount of thermal energy a body gives off or absorbs. Therefore, quantities of heat must be measured by the effects they produce.

Section B: On the Other Hand ...

The following experiment, as best we can tell, was first performed 400 years ago. Try it on your own, with classmates, or at home with your family. Be prepared to describe and explain what happens.

APPENDIX A
SIDE ROADS

1. Fill three open-top containers with water of different temperature as follows: (a) hot, but not scalding; (b) very cold, with or without ice; and (c) lukewarm or room temperature.
2. Immerse one hand in the hot container and the other hand in the cold container.
3. Switch the hand from the hot water to the lukewarm container. Determine whether the water in that container feels "warm" or "cool." Then remove your hand.
4. Switch the other hand from the cold water to the lukewarm container, and again declare the water either "warm" or "cool."
5. Record your findings or predictions, and explain below:
   water in the third container feels: warm / cool (circle one) to the hot hand
   water in the third container feels: warm / cool (circle one) to the chilled hand
   Explanation:

Section C: Actual Factual Factoids

- The *calorie* is defined as the quantity of heat needed to raise the temperature of 1 gram of water 1°C.
- The *calorie* is also defined as a specific number of joules: 4.186.
- The *Calorie* (with a capital C) used in dietary tables by biologists and dietitians to measure the energy value of foods is equal to the *kilocalorie* (1,000 of the calories normally used by physicists).
- The *heat capacity* of a body is the quantity of heat needed to raise its temperature 1°C.
- *Specific heat* is the heat capacity of a material per unit mass.
- Heat capacity = $Q/\Delta T$, where Q is the quantity of heat needed to produce a change (Δ) in the temperature of the body, ΔT.
- Specific heat is calculated by $c = Q/m\Delta T$, where c is specific heat and m is mass.

A

- Water conducts heat away from the body 25 times faster than air does because water has a greater density, therefore a heat capacity 50 times greater than air.
- Staying dry is key to cold weather comfort and survival.
- The metabolism of an organism is very closely tied to temperature. Within the narrow range of temperatures to which the active organism is tolerant, the metabolic rate increases with increasing temperature and decreases with decreasing temperature in a very regular fashion.
- The relationship between metabolic rate and temperature is often expressed in terms of a value called the Q_{10}. This value is a measure of the rate increase for each 10°C rise in temperature. Thus if the rate doubles for each 10°C rise in temperature, the Q_{10} is said to be 2; if the rate triples for each 10°C rise, the Q_{10} is said to be 3; and so forth.
- The exponential nature of the Q_{10} relationship radically affects the activity of poikilothermic ("cold-blooded") organisms as the temperature of their surroundings changes.
- Two classes of animals, mammals and birds, have evolved a mechanism that makes them much less dependent on environmental temperatures and frees them for successful exploitation of more varied habitats. These "warm-blooded," or homeothermic, animals maintain a relatively constant body temperature even when the environmental temperature fluctuates widely.

Part III

APPENDIX A
SIDE ROADS

Homework: Individual Information Search

For each topic you independently research, record below your sources or search engines, keywords, Boolean search strings, and useful URL addresses or bibliographic citations.

Source/Search Engine	Keywords/Search String	Resultant URL/Citation	Notes

Compare your search technique and results with your team members and highlight two strategic search tips:

1.

2.

A

Inquiry Process

The inquiry process is often viewed as a cycle of action that repeats until the investigators reach a satisfying solution. It can be described with seven basic elements:

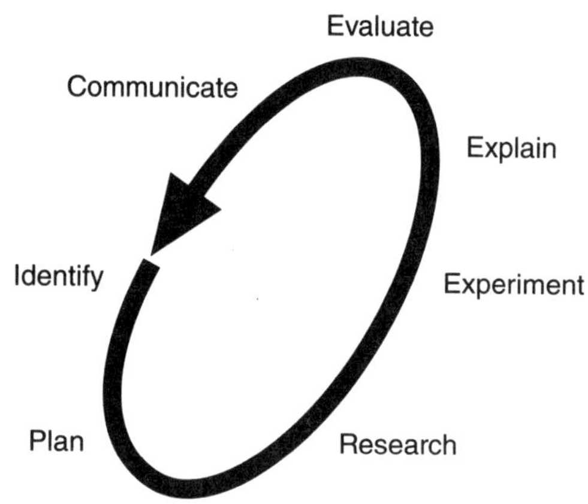

- *Identify* and clarify questions. Understand the issue or problem, and make a testable hypothesis.
- *Plan* appropriate procedures. Brainstorm, draw and write ideas, clarify their ideas, and suggest possible strategies or methods.
- *Research* major concepts. Learn what is known about the situation from sources other than actual investigation, and obtain information from preliminary experiments. Decide what technology, approach, equipment, and safety precautions are useful. Document your experiments and log your data.
- *Experiment.* Use tools and measuring devices to conduct experiments. Use calculators and computers to store and present data.
- *Explain* logical connections. Analyze your data. Formulate explanations using logic and evidence, and possibly by constructing a physical, conceptual, or mathematical model.
- *Evaluate* alternatives. Compare your explanations to current scientific understanding and other plausible models. Identify what needs to be revised, and find the preferred solution.
- *Communicate* new knowledge and methods. Communicate results of your inquiry to your peers and others in the community. Construct a reasoned argument through writing, drawings, and oral presentations. Respond appropriately to critical comments.

APPENDIX A
SIDE ROADS

Questions

Read the following questions, but do not answer them until after your team has experienced working together on the Construct-a-Glove design challenge research activities.

1. Make your own checklist of team activities that correspond to steps in the cycle described above.

2. Create your own version of the inquiry process using words and pathways that fit your team's activity.

3. What shape is your inquiry pathway diagram (circle, spiral, cascade, other)?

4. How and where do the seven steps described above fit within your process description?

A

Homeothermic Regulation

Classification of life forms aids in identification and reveals possible relationships between organisms. There are variations in classification systems, and not all scientists agree on all details. Nevertheless, such systems are useful to organize similar and distinguishing information about life forms. Study of such systems also gives insights into changes that appear over time because of the environment and other factors. In the broadest scientific classification of living things, modern humans belong to the Kingdom Animalia. Further, we belong to the subkingdom Metazoa, section Deuterostomia, phylum Chordata, subphylum Vertebrata, class Mammalia, subclass Theria, order Primates, family Hominidae , genus *Homo,* and species *sapiens.*

Another classification of animals is based on body temperature regulation. Animals obtain energy by cellular respiration, converting chemical energy of food by oxidation of carbohydrates, fats, and proteins into high-energy phosphate bonds of ATP (adenosine triphosphate). Cells are more efficient at converting energy for use than the most elaborate nuclear electric power plant. Even so, 40–60% of the released energy of respiration is in the form of "waste" heat (compared with 65–70% waste heat in producing electricity). Most animals (and all plants) promptly lose this heat energy to their environments. Such animals are called cold-blooded. Scientists use the term *poikilothermic,* meaning "of variable temperature." Only birds and mammals (including humans) have developed insulation (fur, feathers, and fat) and respiratory and circulatory feedback mechanisms to retard the loss of metabolic "waste" heat to the environment. The heat conserved by these warm-blooded animals allows them to maintain a higher body temperature, which sustains a higher metabolic rate (Q_{10}) and increased level of activity. (Refer to Thermal Factoids resource sheet and biology texts for further information.)

Part III

APPENDIX A
SIDE ROADS

Questions

The following questions may help you relate your scientific research and technological development for this project to naturally occurring systems. Insights gained here may stimulate further design innovations.

1. Does the overall metabolic process of animals function more like a furnace or a refrigeration unit? Explain your answer.

2. List several advantages homeotherms gain by maintaining a higher and more constant body temperature than poikilotherms.

3. What added benefits do homeothermic humans get from creating, wearing, and continuously improving insulated gloves?

4. Give two examples of how other animals cope with environmental extremes without gloves.

5. Because humans benefit from gloves, would a frog or a bird benefit from having gloves designed for them?

6. Suggest an event from history that might have been different without the development of gloves.

APPENDIX A
SIDE ROADS

Design Process

The design process is often viewed as a cycle of action that repeats until the designers reach a satisfying solution. It can be described with seven basic elements: identify, create, investigate, choose, implement, evaluate, and communicate.

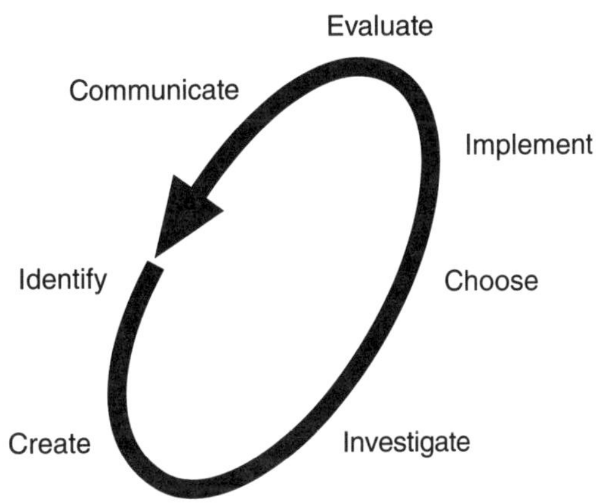

- *Identify* and clarify the situation. Understand the challenge or problem, including the criteria for success and constraints on the design.
- *Create* solutions. Brainstorm, draw and write ideas, and suggest possible strategies or methods.
- *Investigate* possibilities. Learn what is known about the situation, and what technology or approach could be useful. Conduct experiments to test your ideas.
- *Choose* a solution. List the solutions most likely to be successful, and make decisions for how well each solution meets the design challenge or solves the problem.
- *Implement* the design. Learn that a successful design often depends on good fabrication, whether it is a scaled or life-size version of the product.
- *Evaluate* the design. Perform tests to obtain the feedback that informs them about the parts of the design that worked or needed improvement.
- *Communicate* the solution. Present your designs to your peers and others in the community, communicating your ideas through drawings, writing, formal presentations, informal discussions.

A

Questions

After reading about the design process, answer the following questions:

1. Which elements of the process have you already experienced?

2. Which elements have you not yet experienced?

3. Where in the process do you think you are now?

4. What will be your next steps?

Part III

APPENDIX B
Text Reconstruction Exercises

Homeothermic Processes

The paragraphs below describe the homeothermic process and its relation to gloves. To prevent this information from falling into the hands of rival glove manufacturers, the order of the sentences in each paragraph has been jumbled. Your task is to rearrange the sentences in their correct order.

¶ 1

___ Most animals depend on the Sun or on thermal hot springs to keep themselves warm.
___ It involves burning fuel (food) in the body to create heat and using the blood flow as a means of getting that heat distributed evenly.
_2_ A small but a very familiar group of animals are known as warm-blooded because they keep themselves warm.
___ The homeothermic process is the process these animals use to keep themselves at a comfortable temperature.
___ The system is very similar to that in a house that is heated by pumped hot water.
___ The heart and other working muscles produce heat.
___ This heat warms the blood that runs through these warm organs and this warmed blood is then pumped to other areas of the body that may be cooler in temperature.

¶ 2

___ Using the blood flow as a means of conducting heat is a very clever design because this is not what blood was originally invented to do and it is not even now the primary mission of blood flow.
___ Actually the blood in such animals can be hot or cold depending on the surrounding temperature.
___ Many organisms that are not homeothermic use blood for these primary missions; these organisms are called cold-blooded.
___ That primary mission is bringing oxygen and nutrients to the cells and taking away carbon dioxide and waste products.
_5_ Blood can still carry warmth to colder regions of the body, but usually it is from a warm skin to a colder interior.

B

¶ 3

___ In these circumstances, heat can be radiated to the environment by passing overheated blood through the colder surface areas of the body.
___ When an organism is very active it can get too hot.
___ These surface areas are sometimes cooled by the evaporation of sweat.
_5_ The hands, ears, and feet are major areas for heat loss because they have lots of blood flow and a big surface area–to-volume ratio.
___ This evaporation system works just like many of the air-conditioning systems used in houses that have central air-conditioning connected to the same pumped hot water as is used for heating.
___ This big surface area–to-volume ratio acts like the fins on radiators.

¶ 4

___ Hands, ears, and feet are very useful when you are too hot.
___ To protect the rest of the body, the homeothermic system will reduce blood flow to these extremities when body temperature is too low.
___ Eventually, however, the homeothermic system will give up on saving these parts of you and will allow them to freeze and fall off.
___ To outsmart the homeothermic system, people have invented gloves, earmuffs, and socks.
___ This is good in the short term but not a wise choice in the long term.
___ As the temperature of the hands, ears, and feet begins to fall, the body sends short bursts of blood to keep them warm enough to avoid freezing.
_2_ But they become very dangerous when you are too cold.
___ These inventions allow people to live in places like North America and Europe where they probably should never have gone to in the first place.

Part III

APPENDIX B
TEXT RECONSTRUCTION EXERCISES

Heat Energy Transfer

The paragraphs below describe the concept of heat transfer. To prevent this information from falling into the wrong hands, the order of the sentences in each paragraph has been jumbled. Your task is to rearrange the sentences in their correct order.

¶1

___ Heat, which is a form of energy, naturally flows from hot areas to colder areas.
___ The property of materials that determines how fast heat will flow is called thermal conductivity.
___ Other materials such as wood and paper have a relatively low thermal conductivity.
___ If air is trapped so it cannot move around, it becomes a very good thermal insulator.
_6_ Materials with a low thermal conductivity are called thermal insulators.
___ This is why we wear bulky sweaters when we want to be hot.
___ But heat will flow more rapidly in some materials than in others.
___ Many metals, such as copper, have a high thermal conductivity.

¶2

_3_ These new insulators affect how we dress and the design of the buildings in which we live and work.
___ New, lighter, thinner, less expensive thermal insulators are being invented every year.
___ Thermal insulators are very important for clothing and for building construction.

Text Reconstruction Key

The paragraphs below show the correct order of sentences for each text reconstruction exercise. When you hand out the initial homework assignment, ask students to number the sentences in each paragraph so as to put them in the correct order. It is also highly beneficial to ask students to rewrite the paragraphs once they have determined the correct order.

Homeothermic Processes

Paragraph 1

1. Most animals depend on the Sun or on thermal hot springs to keep themselves warm.
2. A small but a very familiar group of animals are known as warm-blooded because they keep themselves warm.
3. The homeothermic process is the process these animals use to keep themselves at a comfortable temperature.
4. It involves burning fuel (food) in the body to create heat and using the blood flow as a means of getting that heat distributed evenly.
5. The system is very similar to that in a house that is heated by pumped hot water.
6. The heart and other working muscles produce heat.
7. This heat warms the blood that runs through these warm organs and this warmed blood is then pumped to other areas of the body that may be cooler in temperature.

Paragraph 2

1. Using the blood flow as a means of conducting heat is a very clever design because this is not what blood was originally invented to do and it is not even now the primary mission of blood flow.

APPENDIX B
TEXT RECONSTRUCTION EXERCISES

2. That primary mission is bringing oxygen and nutrients to the cells and taking away carbon dioxide and waste products.
3. Many organisms that are not homeothermic use blood for these primary missions; these organisms are called cold-blooded.
4. Actually the blood in such animals can be hot or cold depending on the surrounding temperature.
5. Blood can still carry warmth to colder regions of the body, but usually it is from a warm skin to a colder interior.

Paragraph 3

1. When an organism is very active it can get too hot.
2. In these circumstances, heat can be radiated to the environment by passing overheated blood through the colder surface areas of the body.
3. These surface areas are sometimes cooled by the evaporation of sweat.
4. This evaporation system works just like many of the air-conditioning systems used in houses that have central air-conditioning connected to the same pumped hot water as is used for heating.
5. The hands, ears, and feet are major areas for heat loss because they have lots of blood flow and a big surface area–to-volume ratio.
6. This big surface area–to-volume ratio acts like the fins on radiators.

Paragraph 4

1. Hands, ears, and feet are very useful when you are too hot.
2. But they become very dangerous when you are too cold.
3. To protect the rest of the body, the homeothermic system will reduce blood flow to these extremities when body temperature is too low.

B

4. As the temperature of the hands, ears, and feet begins to fall, the body sends short bursts of blood to keep them warm enough to avoid freezing.
5. Eventually, however, the homeothermic system will give up on saving these parts of you and will allow them to freeze and fall off.
6. This is good in the short term but not a wise choice in the long term.
7. To outsmart the homeothermic system, people have invented gloves, earmuffs, and socks.
8. These inventions allow people to live in places like North America and Europe where they probably should never have gone to in the first place.

Heat Energy Transfer

Paragraph 1

1. Heat, which is a form of energy, naturally flows from hot areas to colder areas.
2. But heat will flow more rapidly in some materials than in others.
3. The property of materials that determines how fast heat will flow is called thermal conductivity.
4. Many metals, such as copper, have a high thermal conductivity.
5. Other materials such as wood and paper have a relatively low thermal conductivity.
6. Materials with a low thermal conductivity are called thermal insulators.
7. If air is trapped so it cannot move around, it becomes a very good thermal insulator.
8. This is why we wear bulky sweaters when we want to be hot.

APPENDIX B
TEXT RECONSTRUCTION EXERCISES

Paragraph 2

1. Thermal insulators are very important for clothing and for building construction.
2. New, lighter, thinner, less expensive thermal insulators are being invented every year.
3. These new insulators affect how we dress and the design of the buildings in which we live and work.

APPENDIX C
Sample Answers

Activity 2, p. 209

Identifying Factors and Variables

Glove System Variables	Range of Variation
Example: *Convection seal at wrist*	*no seal* *loose* *tight*
Amount of insulation	*lots > little*
Fit	*tight > bulky*
User-Related Variables	**Range of Variation**
Example: *Starting state of user's hands*	*chilled* *normal* *hot & sweaty*
Level of metabolism	*Text*
Clothing	*Text*
Design Factors	**Constraints**
Example: *Wrist seal pressure*	*Prevent convection without constricting blood circulation*
Insulation thickness	*Bulk reduces movement*
Flexibility of glove material	*Material may tear easily*

Activity 2, p. 210

Team Situation Analysis: Reflection on Your Quick-Build

Most Significant Variables	Suggested Modifications
Example: *Direct heat loss through conduction*	*Slow heat flow with more layers of insulation*
Waterproof materials	*Build out layer*
Soft interior	*Line interior*
Thin, flexible insulators	*Make middle layer*
Sewing seams with waterproof seals/glue	*Construct seams that won't leak*

APPENDIX C SAMPLE ANSWERS

Investigating Heat Transfer and Insulation

Activity 3, p. 221

1. Write down your research protocol for evaluating gloves.

Step 1. Cover glove with thin waterproof glove.

Step 2. Line inside with waterproof glove and fill with room-temperature water, 0.25 liters.

Step 3. Place glove in ice water and measure time for water in glove to drop to 5°C.

"Fair Test" Comparison

Activity 3, p. 222

1. Observe differences between your quick-build insulated glove and a commercial glove that has been designed to be worn in cold weather.

Quick-Build Glove	Commercial Glove
Stiff—hard to move	*Flexible*
Bulky	*Thin*
Falls apart	*Stays together*
Keeps hand warm	*Sort of keeps hand warm*

2. Using your research protocol, compare the insulation effectiveness of the commercial glove to your quick-build product.
Answers will vary.

3. Do the results of your research protocol fit with your understanding of how the two gloves compare?
Protocol shows our gloves as best, but a commercial glove would last longer.

4. Decide whether changes need to be made in your protocol. Is it a fair test?
We need to add durability and flexibility to our protocol.

C

Activity 4, p. 231

Team Feedback

Factors	Suggested Modifications
Insulation leaks out	*Seal cuff*
Glove too bulky	*Reduce insulation*
Seams leak	*Design waterproof seams*
Glove looks awful	*Hire fashion designer*

Activity 4, p. 232

Reflections on Design

1. What is the one feature of your insulated glove system about which you are most proud?
Our glove is better insulated than the commercial glove.

2. Describe the concerns you have with the present design.
The glove drips whipped cream.

3. Describe in a paragraph or two the approach your team took in the designing and building of your prototype. Include a problem your team encountered with the performance tests and how you went about solving it.
We could not think of a good insulator that was cheap, flexible, and wouldn't flatten down. We thought of using peanut butter, but that was too sticky. Then we decided to use whipped cream.

Part III

Glossary

baseline data: data taken to determine conditions before an experiment is started

bioregulation: a biological process that controls the value of a variable. For example, bioregulation keeps body temperature constant in homeothermic animals

brainstorming: a group problem-solving technique which involves the spontaneous contribution of ideas from all members of the group

calorie: the amount of heat needed to increase one gram of water by 1°C

cellular respiration: the conversion within the cell of nutrients (such as sugar molecules) into chemical energy, by reacting the food with oxygen (O_2); by-products are carbon dioxide and water

conduction: the flow of thermal energy through a substance from a higher- to a lower-temperature region

constraints: restrictions imposed on a system or process

convection: the movement of heat through a fluid (such as liquid or gas); for example, the free or forced movement of warm air throughout a room

criterion: a standard on which a judgment or decision may be based

design factors: a set of variables and constraints considered critical to a successful design

dexterity: skill and ease in using the hands

"fair test" comparison: a comparison among two or more designs in which no design is given an advantage

friction: resistance to sliding, a property of the interaction between two solid bodies in contact

heat: the amount of internal kinetic energy of atoms and molecules that flows from a warmer to a cooler environment (see temperature)

heat capacity: the amount of heat required (measured in joules or calories) to raise a unit of mass (measured in

GLOSSARY

grams or kilograms) by 1°C; for example, raising 1 gram of water by 1°C requires 1 calorie of energy

homeotherm: an organism that maintains a constant internal body temperature (see poikilotherm)

insulation: material used to prevent transfer of heat

iteration: the action of repeating a process

joule: a metric unit for work that is the force of 1 newton acting over a distance of 1 meter, where the force and the distance moved lie in the same direction; named after James Prescott Joule, a 19th-century British physicist

kinetic: of or relating to the motion of material bodies and the forces and energy with which that motion is associated

metabolism: all the physical and chemical processes by which living substance is produced and maintained; the transformations by which energy is made available for use by an organism

newton: a metric measurement of force based on 1 kilogram of mass experiencing acceleration of 1 meter per second squared

parameter: any of a set of physical properties whose values determine the characteristics of something

performance test: a test to determine how well a design accomplishes an intended purpose

poikilotherm: an organism with a variable body temperature

prospectus: a preliminary printed statement that describes a product or an enterprise; the prospectus is distributed to prospective participants, investors, or buyers

protocol: the plan of a scientific experiment or treatment

prototype: an original model; the first full-scale and (usually) functional form of a new type or design of a construction

Q_{10}: a value representing the relationship between metabolic rate and temperature;

GLOSSARY

measures the metabolic rate increase for each 10°C rise in temperature (e.g., if the rate doubles for each 10°C rise in temperature, the Q_{10} is said to be 2)

radiation: a type of heat transfer through exposure to a series of electromagnetic waves, such as infrared waves.

range of variation: the least and greatest values found for a specific variable

specific heat: the quantity of heat needed to produce a change in the temperature of a specific body mass ($Q/m\Delta T$)

specifications: the precise details of an invention, plan, or proposal

system: a regularly interacting or interdependent group of items forming a unified whole

temperature: a scale for measuring thermal energy by showing how warm or cold an object is relative to something else (such as the freezing and boiling points of water)

thermodynamics: the study of heat and how heat energy is transformed from one form of energy to another

variable: an object or quality of changeable value

vasoconstriction: narrowing of the blood vessels

vasodilatation: widening of the blood vessels

Suggested Readings

Byalko, A. 1997. Hands-on (or-off?) science: Why is thermal sensitivity a touchy subject? *Quantum/Feature* Nov/Dec: 4–8.

Curtis, R. 1995. *Outdoor action guide to hypothermia and cold weather injuries.* Princeton University Outdoor Action Program. *www.princeton.edu/~oa/hypocold.html*

Custer, R. L. 1996. Rubrics: An authentic assessment tool for technology education. *The Technology Teacher* Dec: 27–37.

Keys, C. W. 1994. The development of scientific reasoning skills in conjunction with collaborative writing assignments: An interpretive study of six ninth grade students. *Journal of Research in Science Teaching* 31 (9): 1003–1022.

McCormick, R., S. Hennessy, and P. Murphy. 1993. Problem-solving processes in technology education. Paper presented at the 55th Annual Conference of the ITEA, Charlotte, North Carolina.

Petroski, H. 1996. *Invention by design: How engineers get from thought to thing.* Cambridge, MA: Harvard University Press.

Renner, J. W., M. R. Abraham, and H. H. Birnie. 1983. Sequencing language and activities in teaching high school physics: A report to the National Science Foundation. Norman: University of Oklahoma.

Richmond, G., and J. Striley. 1996. Making meaning in classroom: Social processes in small-group discourse and scientific knowledge building. *Journal of Research in Science Teaching* 33 (8): 839–858.

Roberts, N., D. Andersen, R. Deal, M. Garet, and W. Shaffer. 1983. *Introduction to computer simulation: A system dynamics modeling approach.* Productivity Press.

Roth, W. M. 1994. Experimenting in a constructivist high school physics laboratory. *Journal of Research in Science Teaching* 31 (2): 197–223.

Roth, W. M. 1995. *Authentic school science: Knowing and learning in open-inquiry science laboratories. Science and Technology Education Library*. Boston: Kluwer Academic Publishers.

Schuring, D. J. 1977. *Scale models in engineering*. New York: Pergamon Press.

Tracy, K. 1991. *Understanding face to face interaction: Issues linking goals and discourse*. Hillsdale, NJ: Erlbaum Associates.

Wilson, M., and K. Draney. 1996. Mapping student progress with embedded assessments. Paper presented at the annual meeting of the American Educational Research Association, New York.

Part IV

CONSTRUCT a GREENHOUSE

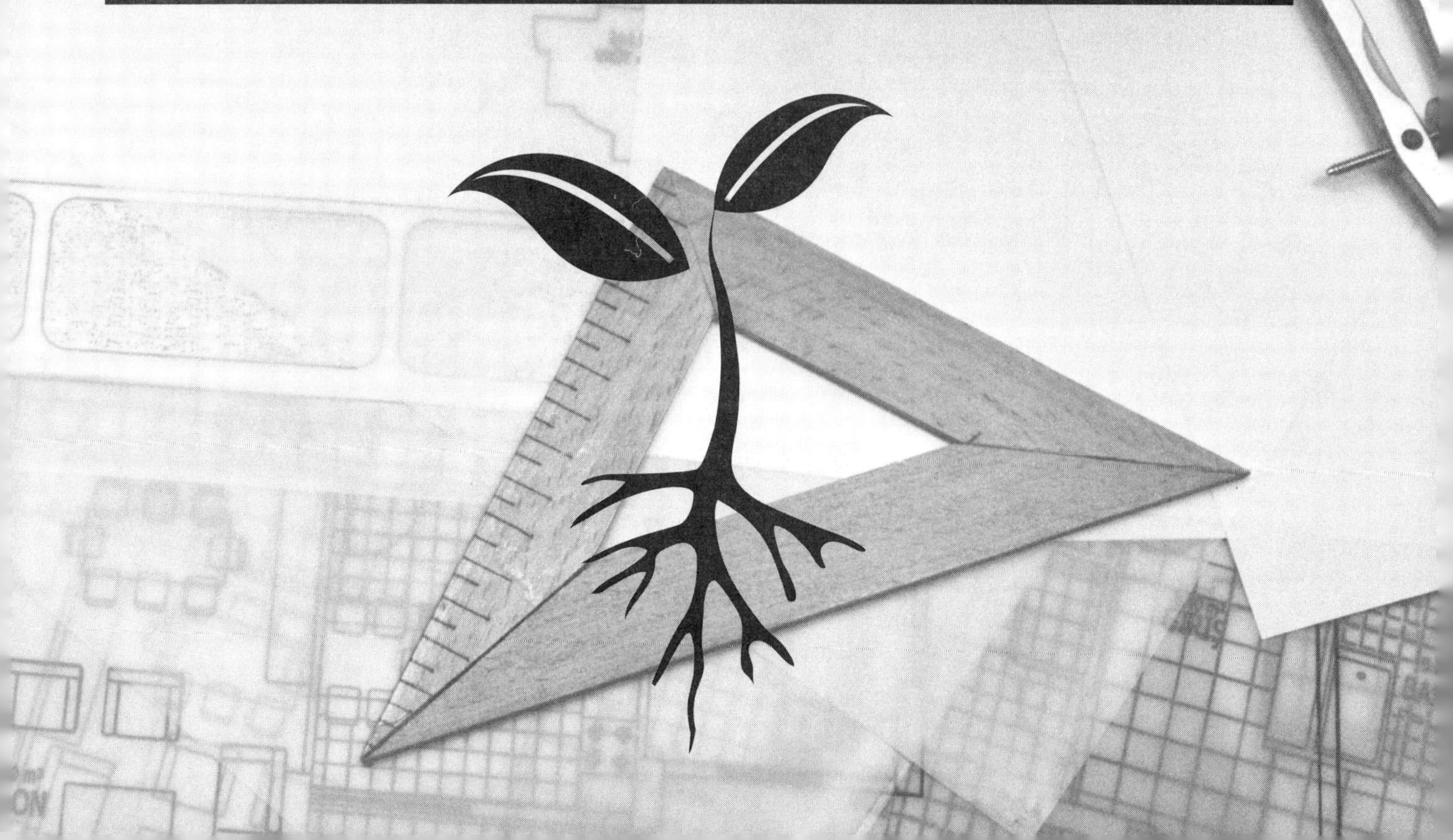

Part IV

Introduction

Integrating Science and Technology

Construct-a-Greenhouse is aligned with the National Science Education Standards for process and content in physical science, biology, and technology, as shown in the Standards and Benchmarks Connections table later in this section. Through a variety of hands-on design activities, students engage in the iterative processes of scientific inquiry and technological design to develop conceptual understanding of heat energy transfer, photosynthesis, plant metabolism, thermal regulation, and feedback control.

Schedule and Cost

You need at least 13 days to complete the core unit. If you have time, you may choose to cycle through the research and development phases several times, either to explore science concepts in greater detail or to produce a state-of-the-art greenhouse.

Students develop models of their designs, build the models, conduct further investigations, analyze their data, redesign if necessary, and communicate their results both orally and in writing. Students are given the Design Brief and instructions for making a quick-build greenhouse (approximately three days). During the research and development phases (at least seven days), students identify relevant variables, design and conduct experiments, and generate possible solutions to their design obstacles. In this way, students come to think of research and development as a repeating cycle. This unit concludes with students communicating their product and their scientific argument to their classmates (three days).

Construct-a-Greenhouse works particularly well in a multidisciplinary course, or taught as a team effort involving faculty in physics, technology, and biology. The biological demands of this project vary with geographic location, altitude, time of year, and available solar exposure. Students should research the time requirement for each growth phase of the selected plant and should scale their prototype growth enclosure to fit with the calendar time allotted for the project.

Cost of consumables can be kept at about $7 per student, if materials are carefully controlled. If you decide to undertake full-scale construction for the prototype-building phase, this will, of course, dramatically increase costs.

Key Ideas

Each key science idea used in Construct-a-Greenhouse is covered in a text reconstruction exercise in Appendix B (p. 361). The first exercise, on energy transfer, involves a very simple reconstruction and is intended as an introduction to text reconstruction. The other exercises are more difficult.

Transformation and Transfer of Energy

Energy can change from one form to another in a controlled physical environment just as it does in the living world. In Construct-a-Greenhouse, students build light trap systems, in which light is changed to heat. They then learn to use the principles of heat transfer to design and build a system for growing a plant of a particular species.

Repurposing and Trade-Offs

The values of any technology may be different at different points in time. To start their seeds, students must engineer the growth enclosure so that it will convert light to heat. Once the seeds sprout, the design emphasis shifts toward maximizing light intensity. Students *repurpose* the physical system (the greenhouse) to provide an optimal environment for the biological system (the plant) as it grows through different stages. Also, students must make numerous decisions involving trade-offs as they encounter cost constraints, construction alternatives, and other design factors.

Plant Growth and Stages of Development

The needs of a plant vary as it goes through different stages of its life cycle. These stages are germination, seedling, leaf, stem, and root growth, and fruit development.

Three important factors in the germination of nearly all seeds are water, oxygen, and temperature. After a seed germinates, the growing seedling requires carbon dioxide, water, and light to support photosynthesis. To produce new leaves, stems, roots, and fruit, plants must obtain additional essential nutrients from the soil in which they grow. Most weight gain comes from the process of photosynthesis, which converts carbon, oxygen, and hydrogen in air and water into plant mass. Deficient factors can limit growth, as can an excessive supply.

Student Portfolios

The following items can be accumulated in portfolios for summative assessment:

- Pretest: Snapshot of Understanding
- Initial questions: Design Brief
- Individual information search
- Sketch of quick-build
- Brainstorming record
- List of variables
- Research and results
- Group process description: Inquiry Process
- Group process description: Design Process
- Prototype demonstration notes
- Group summary documentation
- Snapshot of Understanding (posttest and self-assessment)

INTRODUCTION

Standards and Benchmarks Connections

Task	Source
Students identify and describe variables that affect functioning of a simple, passive solar collection system to store and convert radiant energy for specific purposes. **Standard/Benchmark:** Energy Transformations; Motion; Flow of Matter and Energy; Physical Science Content Standard B; Earth and Space Science	AAAS 9–12, NSES 9–12
Students identify alternatives in system components and configuration, and develop experimental proposals to investigate them. **Standard/Benchmark:** Design and Systems	AAAS 9–12
Students interpret scale drawings and interpret and draw three-dimensional objects. **Standard/Benchmark:** Communication Skills; Geometry from a Synthetic Perspective	AAAS 9–12, NCTM 7
Students learn critical components of scientific inquiry by reflecting on concepts that guide the inquiry, and by establishing an adequate knowledge base to support their investigation. **Standard/Benchmark:** Scientific Worldview; Science as Inquiry	AAAS 9–12, NSES 9–12
Students research and organize information about requirements for optimal growth of the Atlantic Giant pumpkin, limiting factors, and tolerance. **Standard/Benchmark:** Flow of Matter and Energy; Nature of Technology; Science as Inquiry	AAAS 9–12, ITEA II, NSES 9–12
Students research the scientific principles that support some "rules of thumb" for building a passive solar energy trap. **Standard/Benchmark:** Energy Transformations; Nature of Technology; Science as Inquiry; Physical Science	AAAS 9–12, ITEA I and III, NSES 9–12
Students keep a laboratory notebook/design log to document their work and to reflect on their work. **Standard/Benchmark:** Computation and Estimation; Science as Inquiry	AAAS 9–12, ITEA III, NSES 9–12
Students practice design process as an iterative cycle of multidisciplinary activities including research, choosing between alternatives, drawing, building, testing, and evaluating. **Standard/Benchmark:** Design and Systems; Nature of Technology	AAAS 9–12, ITEA I
Students identify relevant materials properties and make choices based on constraints of cost, suitability to criteria, and availability. **Standard/Benchmark:** Nature of Technology	ITEA I
Students use previously researched rules of thumb to mathematically scale aspects of their design. **Standard/Benchmark:** Scale; Mathematics as Problem Solving	AAAS 9–12, NCTM 1

INTRODUCTION

Task	Source
Students identify alternatives in system components and configuration, and build a prototype for performance testing. **Standard/Benchmark:** Energy Transformations; Nature of Technology; Science and Technology	AAAS 9–12, ITEA II and III, NSES 9–12
Students reflect and write recommendations based on their design tests. **Standard/Benchmark:** Communication Skills; Nature of Technology	AAAS 9–12, ITEA II
Students give an oral presentation on their prototype design specifications, its operating procedures, and the quantitative and qualitative plant growth results obtained. **Standard/Benchmark:** Communication Skills; Nature of Technology	AAAS 9–12, ITEA I
Students articulate principles of science employed in passive solar greenhouse operation. **Standard/Benchmark:** Critical-Response Skills; Technology and Society; Life Science	AAAS 9–12, ITEA VII, NSES 9–12
Students self-assess their learning by comparing pre- and post-Snapshots of Understanding. **Standard/Benchmark:** Issues in Technology; Science, Technology, and Society	AAAS 9–12, NCSS VIII

Source Key:

AAAS = American Association for the Advancement of Science. 1993. *Project 2061: Benchmarks for science literacy.* New York: Oxford University Press.

ITEA = International Technology Education Association. 1996. *Technology for all Americans: A rationale and structure for the study of technology.* Reston, VA: ITEA.

NCSS = Task Force on Social Studies Teacher Education Standards. 1997. *National standards for social studies teachers.* Washington, DC: National Council for the Social Studies.

NCTM = National Council for Teachers of Mathematics. 1991. *Professional standards for teaching mathematics.* Reston, VA: NCTM.

NSES = National Research Council. 1996. *National science education standards.* Washington DC: National Academies Press.

Part IV

INTRODUCTION

Course Outline

Introduction

- Overview: Greenhouse Design Brief
- Snapshot of Understanding

Quick-Build

- Overview: Quick-Build Greenhouse
- Quick-Build Greenhouse Specifications
- Qualitative Observations
- Seed Starting

Research

- Overview: Research
- Species Specifications
- Tolerance and Limiting Factors
- Identifying Variables and Making Measurements
- What We Do With the Data
- Rules of Thumb

Development

- Overview: Development
- Criteria and Materials
- Scaling and Costing
- Design Drawing
- Prototype Construction
- Performance Tests
- Proof-of-Design Performance Evaluation

Communication

- Overview: Communication
- Presentation
- Frequently Asked Questions
- Reflection and Recommendations
- Estimation: How Much Does Your Pumpkin Weigh?
- Snapshot of Understanding

DAY 1

ACTIVITY 1
Greenhouse Design Brief

OVERVIEW: GREENHOUSE DESIGN BRIEF

In this unit, you will be designing and building an engineered environment for growing a giant vegetable. You will need to do some research on what the optimal conditions are for seeds to sprout, for seedlings to thrive, and for the vegetable to grow to maximum size. You will also need to do research on desirable sizes, shapes, and materials for your greenhouse.

Design Challenge

As a member of a development team, design and build an environment adaptable to changing heat, light, humidity, and space requirements for the progressive growth stages of a giant pumpkin (or other specified fruit or vegetable).

Scope of Work

- *Quick-Build:* Build a simple greenhouse according to specifications.
- *Research:* Conduct experiments to collect, record, and communicate baseline data. Explore science concepts related to the functioning of the greenhouse.
- *Development:* Modify initial greenhouse or design anew to accommodate specific requirements of your plant at different growth stages.
- *Communication:* Summarize orally the outcomes of your effort, respond to questions, and provide written advice to assist others in achieving success.

Part IV

ACTIVITY 1
GREENHOUSE DESIGN BRIEF

SNAPSHOT OF UNDERSTANDING
What I Already Know About Light Absorption and Energy Conversion

This activity will help you review what you already know about the processes of absorption and conversion of light energy to heat. It will also help you think about certain features to consider in your design of a greenhouse that you can optimize for changing plant growth requirements. The activity involves a short experiment in which you will assemble an "insta-build" greenhouse according to instructions; measure temperature changes as an indication of energy absorption, conversion, and transfer over a brief time interval; and answer four questions.

Materials

Part	Type	Size or Quantity
Box	Cardboard, foam, or wood	20 cm × 13 cm × 6 cm
Clear cover	Transparent plastic/film, glass, or Plexiglas	single sheet, minimum size 22 cm × 15 cm × 0.025 mm
Thermal mass	Aluminum soda can	1
Spray paint	Quick-drying, black, flat or matte finish	1 can
Insulation	Bubble wrap or newspaper	1 sheet
Glue stick	Quick-bonding	1
Thermometers	Alcohol, 0–50°C	2 (one small enough to fit in box)
Light source	Desk lamp with 60 W lightbulb (or equivalent)	1
Tape		

Assembly Instructions

1. Assemble your small greenhouse according to the perspective drawing shown below.
2. Crush soda can flat and paint it black.
3. Tape one thermometer to an inside wall and the other on an outside wall of the insta-build.
4. Glue or tape transparent film to the box edges, sealing the edges against air leakage.

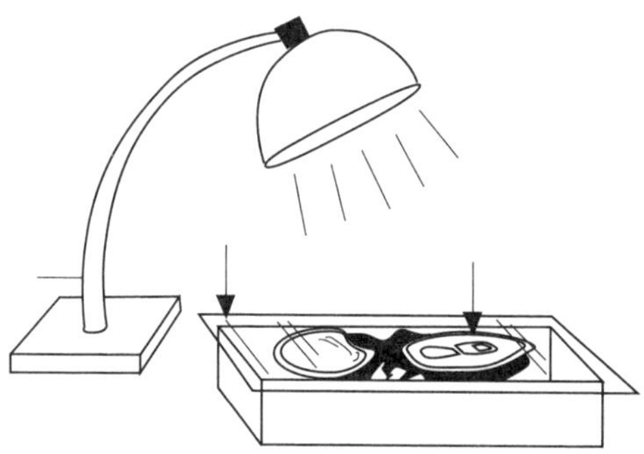

Data Collection

- Position your small greenhouse so that the absorber surface (flattened soda can) is 90° and 15–20 cm from the 60 W lightbulb of a desk lamp.
- Read both thermometers just before you turn the light on, record your readings, and note the time.
- Answer questions 1–3 on page 289 (you have 20 minutes before taking a second thermometer reading).
- Take a second reading on each of the thermometers about 20 minutes after your first reading. Record your data in your portfolio notebook, then answer question 4.

Part IV

ACTIVITY 1
GREENHOUSE DESIGN BRIEF

Questions

1. When objects of different temperatures are in contact, heat flows from the warmer to the cooler, eventually reaching a common temperature. Describe in one or two sentences each of the listed key processes of temperature equalizing.

Conduction:

Convection:

Radiation:

SCILINKS THE WORLD'S A CLICK AWAY

Topic: conduction/convection
Go to: *www.scilinks.org*
Code: CAH01

Topic: radiation
Go to: *www.scilinks.org*
Code: CAH02

Topic: photosynthesis
Go to: *www.scilinks.org*
Code: CAH03

2. Why does a good absorber of radiant energy appear black?

3. What does it mean to say that the "greenhouse effect" is like a one-way valve?

Read the thermometers and record your data.

4. In about five sentences, describe what you have learned from your observations and measurements about visible light absorption and its conversion to other forms of energy. (Remember—this is not a test; there are no right or wrong answers.)

DAY 1

ACTIVITY 1 Teacher Pages

OVERVIEW: GREENHOUSE DESIGN BRIEF

Give students the Construct-a-Greenhouse design challenge and the Snapshot of Understanding. The Snapshot includes a mini-experiment called an "insta-build," which is best done in small groups of two or more students, depending on the space available to set up. This Snapshot is a performance assessment of student understanding. It uses construction and measurement activities combined with short, written questions.

Design Challenge

Have students read the design challenge. Advise students that they will use processes in both technological design and scientific inquiry to meet the challenge. They will then collect data on an insta-build experimental greenhouse, design and test an improved engineered environment, and finally present and defend their team's activities as proof of meeting the challenge.

Snapshot of Understanding

In a single class session, students work in pairs, following instructions and reading drawings to construct a very simple "insta-build" greenhouse. They expose the greenhouse to either solar or incandescent radiation and record temperature increases after 20 minutes. While waiting for 20 minutes to pass, they answer three of four *Snapshot* questions. After completing measurements, students write short answers to questions about their observations.

Preparation

- Read and become familiar with the entire unit.
- Photocopy student activity sheets for distribution.
- Obtain "insta-build" construction and testing materials.
- Tailor instructions and assembly drawings to materials obtained.
- Define your assessment system with a clear, simple description.

ACTIVITY 1
TEACHER PAGES

Materials

For Each Student:

- Student Activity Sheets
    - » Course Outline
    - » Overview: Greenhouse Design Brief
    - » Snapshot of Understanding
- Ring, Pocket, or Folio Binder (student supplied; for keeping student activity sheets, notes, and drawings for reference and presentation)

For Each Team:

Insta-Build Materials

- shallow cardboard, foam, wood, or otherwise nonconductive box, open on one side—a bit larger in length and width than a crushed soft drink can; 20 cm × 13 cm × 6 cm
- clear transparent sheet, such as overhead transparency, glass, or Plexiglas (do not use plastic food wrap; it is too thin); minimum size 22 cm × 15 cm × 0.025 mm
- aluminum soft drink can
- quick-drying black spray paint, flat or matte finish
- sheet insulation material, such as bubble wrap or newspaper
- quick-bonding glue stick
- two alcohol thermometers (0–50°C)
- light source: direct sunlight or lamp with 60 W lightbulb (or equivalent)
- tape
- scissors
- clock or watch

- tape
- pencil and paper
- disposable camera, with flash (or digital or instant camera)

Time Requirement

This activity requires one class session. Allow about 35 minutes for students to complete the Snapshot of Understanding.

Teaching Suggestions

Hand out the Greenhouse Design Brief student activity sheets. Ask students to keep these and future sheets together and to bring them to the classroom with other notes to serve as a record and reference for daily activity (and assessment) in the unit. Divide the class into pairs or teams of three. Advise students that they will use processes of technological design and scientific inquiry together; that other teams will critique their prototype with respect to the challenge criteria; and that they must each document their activity to contribute effectively to both the final team presentation and their individual portfolio assessment.

Students are also required to document their activity in a laboratory journal in order to contribute effectively to the final team presentation and to enhance their individual portfolios. Be clear on your rubrics for assessing their work. Indicate which activities will be individually graded and which will be given a team score. Be prepared to justify team scoring if some students (or parents) are not used to the idea.

Preassessment

Hand out the Snapshot of Understanding. Emphasize that it is not a test and that students will not be graded on this activity. The purpose of the Snapshot is self-diagnostic—to find out what students know initially about the key science and technology learning objectives of Construct-a-Greenhouse.

An inventory of students' prior knowledge is an important teaching and learning tool. Not only does the inventory help guide students toward the concepts they need

ACTIVITY 1
TEACHER PAGES

to learn the most, but it also prepares them to accept new information in a manner that ties meaningfully to what they already know. At the end of Construct-a-Greenhouse students will be able to compare these answers with their answers in another Snapshot of Understanding.

Collect and retain the Snapshots after students complete them.

Issuing the Construct-a-Greenhouse Challenge

Refer the class to the Greenhouse Design Brief and discuss the challenge statement. Indicate how far you will be going with the challenge. Will the project stop in a few weeks when the seeds have sprouted, or will you be following progress for several months until you can harvest giant pumpkins? If the former, you may want to donate the seedlings to the care of a class of younger students, day care center, or community garden. This will give the project a greater sense of purpose. To get ideas for expanding the challenge, look through a greenhouse supply catalog (e.g., Charley's Greenhouse, *www.charleysgreenhouse.com*).

This unit contains directions for growing Atlantic Giant pumpkins. If you use another type of seed, adjust the instructions according to the seed package directions.

2 DAYS

ACTIVITY 2
Quick-Build Greenhouse

OVERVIEW: QUICK-BUILD GREENHOUSE

The quick-build greenhouse is a simple enclosure for absorbing, converting, and accumulating light energy as heat. You can build the greenhouse quickly, within one class period. You will use specified materials to build this enclosure, following technical instructions. You will then use the quick-build to make some baseline measurements to inform your research and development efforts in later activities.

The practice of building to specifications ensures that everyone starts in the same way, so it is important that you construct your quick-build according to the technical instructions and drawings provided. You will be assessed later on the improvements you can make to the quick-build's baseline performance.

Quick-Build Greenhouse Specifications

Materials for Each Team:

- two black or white vinyl-clad 3 in. ring binders
- duct tape, 2 in. width
- clear plastic film, 30 cm × 40 cm, at least 0.05 mm thick
- alcohol thermometer (0–50°C)
- 60 W adjustable desk lamp (gooseneck, clamp or tension arm)
- scissors
- ruler
- two plastic planting pots, 4-inch square or equivalent size round pots
- two coffee filters, paper towels, or other liner for the pots
- potting soil, sterile, light tilth mix, approximately 1 pint
- two to six seeds (provided by the teacher)
- water (approximately 250 ml per pot)

ACTIVITY 2
QUICK-BUILD GREENHOUSE

Instructions

1. Review the quick-build greenhouse schematic.
2. Cut one cover flap off each ring binder at its vinyl seam. Take care not to expose the cardboard edge; if exposed by mistake, seal the edge with moisture-proof (duct) tape.
3. Holding the parts in place (one binder spine with attached cover forming partial top and back, the other binder cover with attached spine forming partial front and bottom, the detached covers forming side walls), mark end points for folds where side walls meet; cut edges of top and front panels. Refer to schematic for visual aid.
4. Connect end point marks with straight edge and bend (or cut off) the corner of each separated cover flap to make side wall edge slant as shown.
5. Use duct tape to fasten all adjoining edges together in the configuration indicated on the diagram.
6. Tape the thermometer in the binder rings at top of greenhouse.
7. Cut the clear plastic sheet to fit opening with ample overlap on all sides, and tape it at the top only.

Schematic

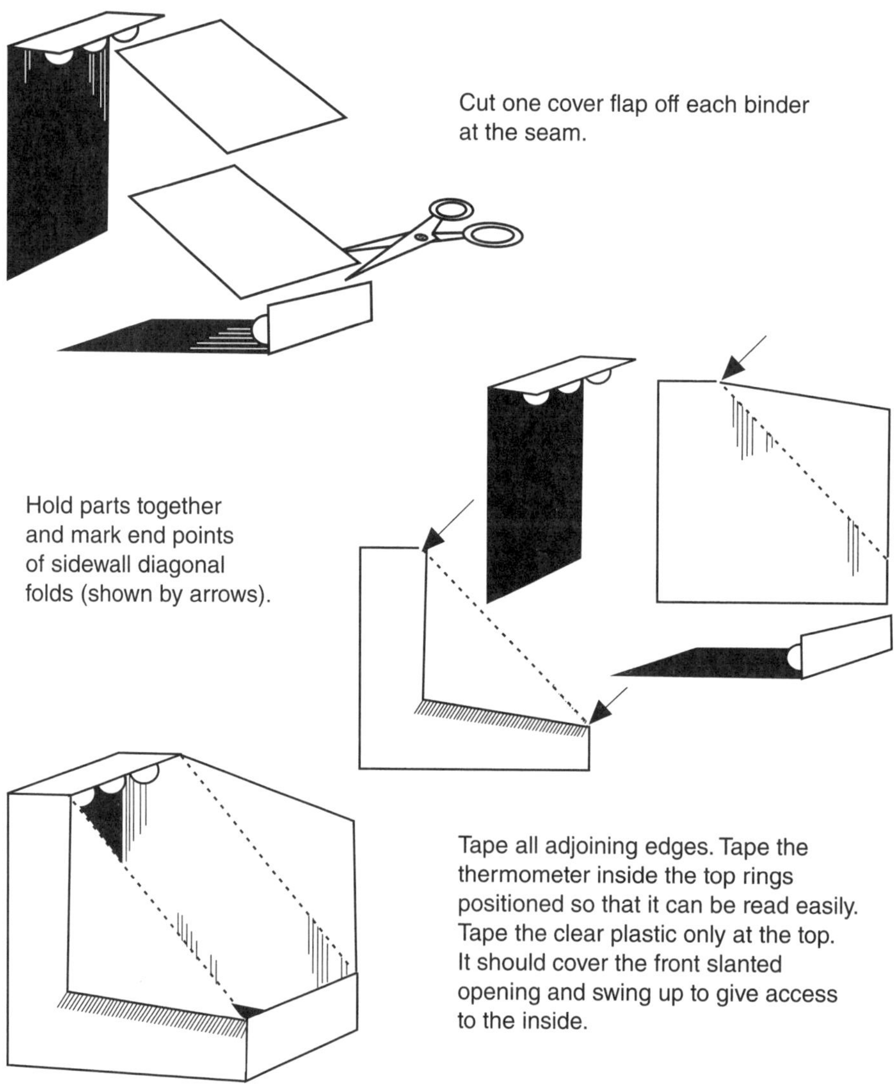
Cut one cover flap off each binder at the seam.
Hold parts together and mark end points of sidewall diagonal folds (shown by arrows).
Tape all adjoining edges. Tape the thermometer inside the top rings positioned so that it can be read easily. Tape the clear plastic only at the top. It should cover the front slanted opening and swing up to give access to the inside.

ACTIVITY 2
QUICK-BUILD GREENHOUSE

Qualitative Observations

Keeping a Journal

Not only is a journal an important tool for inventors, it is particularly useful for biological research. Unlike many physical systems, a living, growing, biological system cannot be measured easily or evaluated quantitatively. Biological systems seldom provide the observer with simple correspondences among various components within a system. Growth occurs in stages and takes varying amounts of time. One way to study a biological system is to keep careful descriptive records of procedures and observations over an extended period of time.

Keep a laboratory journal. The purpose of keeping a journal is to record as much detail as is necessary, as concisely as possible, so that you or other researchers can repeat the experiment later. It is also important to put dates on your records. It is usually unnecessary to write flowery prose when you write in your notebook or journal.

Laboratory notebooks or journals are often used in the patent process to establish the time during which certain results were achieved, for the purpose of protecting patent rights to processes or products.

Observing the Biological System

You will shortly begin preparing your greenhouse for sprouting seeds. Listed below are the kinds of information you will want to record in your journal as you are preparing the seedbed.

SCILINKS. THE WORLD'S A CLICK AWAY

Topic: plant life span
Go to: *www.scilinks.org*
Code: CAH04

Topic: primary/secondary growth
Go to: *www.scilinks.org*
Code: CAH05

Observing Seed Germination

Start your journal entry on the day that you begin preparing an environment for starting your seeds. Enter your thoughts (including why you choose to do one thing and not another) and describe your procedure. Record the following information:

- soil characteristics, including
    - mixture of organic and inorganic materials
    - soil texture when dry and soil texture when saturated with water
    - how well the soil holds moisture
- how far below the surface the seeds were planted
- soil temperature and how the temperature is to be maintained

- how watering is to be done, how often it is done, and the amount of water added
- number of days until a seedling emerges

Observing the Seedling

Note the day that you first see seed leaves emerge from the soil in your laboratory journal as Day 1 of the seedling stage. (Seed leaves are attached to the outer covering of the seed, so you will probably see the whole seed emerging.) Then, on each subsequent day, enter in your observation log all relevant information, such as

- height of seedling
- number of seed leaves
- number of days until true leaf appears
- size of leaves (dimensions, leaf area, total leaf area)
- leaf surface reflectance (shiny or dull)
- height: diameter ratio
- general appearance (stoutness, texture, coloration, etc.)

Recording the Physical Conditions

For each journal entry, note and record all pertinent conditions, such as the following:

- start and end times of light periods
- maximum and minimum air temperatures (internal and external)
- soil temperature
- relative humidity
- time of day of watering and amount of water (in milliliters)

ACTIVITY 2
QUICK-BUILD GREENHOUSE

Seed Starting

To provide optimum conditions for the process of germination, most growers start their seeds in small containers and later transplant the seedlings into the garden. This process is called *indirect seeding*—when seeds germinate in a location different from that of the final growing area.

At the peak of the fruit growth stage of the Atlantic Giant pumpkin, weight gain can be as much as 25 pounds per day. This means that every extra day added at the beginning of the growing season could mean a 25-pound difference at the time of harvest for contest weighing. To gain this advantage, the greenhouse you construct will be designed to optimize conditions for rapid germination.

Materials

- your team's quick-build greenhouse
- two or three plastic planting pots, 4-inch square or equivalent size round pot (one used for control)
- pot saucers or dishes to hold drained water
- coffee filters, paper towels, or other pot liners
- potting soil: sterile, light tilth mix (approximately 1 pint)
- Atlantic Giant pumpkin (or other) seeds (two to six)
- water (about 250 ml per pot)
- 60 W adjustable lamp (gooseneck, clamp, or tension arm)
- masking tape, sticks, pot labels, ruler

Starting the Atlantic Giant Pumpkin Seed

1. Fill and lightly pack the 4″ pots with potting soil.
2. Saturate the planting mixture with water, and place the pots in drain dishes.
3. Plant one seed in each pot, 2–3 cm below the soil surface, with the pointed end of the seed down.
4. Put one or two seeded pots inside your quick-build greenhouse. How will you keep track of the seedlings later on?
5. Position your quick-build for exposure such that the absorber surface is 90° to the light source (desk lamp) and 15–20 cm from the 60 W lightbulb.
6. Position the remaining pot(s) outside the quick-build, about 30.5 cm from the light source but well shielded from that source.
7. Read both the inside and the outside thermometers just before you turn the light on, and record your readings in your laboratory journal.
8. Note the starting time.

Over the next few days:

1. Leave the light on for the entire germination stage.
2. If you have not done so yet, start your laboratory journal *today*.
3. About three times a day, if possible during school hours, team members should observe the quick-build temperatures and the status of the pots. Record the following information in your observation log and consolidate individual readings daily into a team record sheet:
    - Date and time
    - Thermometer readings on both the inside and the outside
    - Is there any condensation in the system?
    - Are there any signs the seedlings will break out of the soil soon?

ACTIVITY 2
QUICK-BUILD GREENHOUSE

4. The ideal temperature range for germination is 24–27°C. What would you do if the temperature you observe is different from the ideal temperature range? In your laboratory journal, note any steps taken to correct deviations from the ideal temperature.
5. Specifically note the amount and location of any condensation within the greenhouse. If you notice condensation on an inside surface of the quick-build, what does it mean? Will it affect the optimal condition for germination? If so, how will you correct the situation? In your laboratory journal, note what corrective action, if any, is taken.
6. If you notice the soil bulging (a sign that the seedling will be breaking out within a few hours), let the rest of the team know, and keep a close watch for the appearance of the seedling. Record the date and time of the first indication of imminent or actual seedling emergence.

2 DAYS

ACTIVITY 2 Teacher Pages

OVERVIEW: QUICK-BUILD GREENHOUSE

In a single, fast-paced class session, students work in groups of three, following instructions to construct a quick-build greenhouse. Following instructions in the Seed Starting activity sheet, student teams prepare and plant seedbeds. The activities that follow must be spread out over several days and can be conducted either during parts of Activity 3 or during some other activity of your own design. The total time needed depends on the kind of seed used and on greenhouse temperatures.

Because Construct-a-Greenhouse stresses scientific inquiry, much of the quick-build activity involves careful implementation of precise instructions. Following and writing precise instructions are important elements of inquiry; they are required to ensure a standard starting point from which progress is assessed. Students may need your help to understand when they are expected to follow instructions precisely and when they should innovate. In Construct-a-Greenhouse, the first two activities tend to be more formulaic than the remaining three, which require more independent thought.

Preparation

- Obtain quick-build construction and testing materials.
- Tailor quick-build instructions to materials obtained.
- Consider noise, safety, and access to water and light in choice of work sites.
- Designate and arrange assembly and testing areas for the quick-build.
- Organize materials for easy access.
- Determine student team formation strategies.
- Contact related workplace representatives, such as nursery operators, for establishing context and relevance of the lesson.
- Preview library and internet resources.

ACTIVITY 2 TEACHER PAGES

Materials

For Each Student:

- Student Activity Sheets
    - » Overview: Quick-Build Greenhouse
    - » Quick-Build Greenhouse Specifications
    - » Qualitative Observations
    - » Seed Starting

For Each Team:

- two black or white vinyl-clad 3 in. ring binders
- duct tape, 2 in. width
- clear plastic film, 30 cm × 40 cm, at least 0.05 mm thick
- alcohol thermometer (0–50°C)
- 60 W adjustable desk lamp (goose neck, clamp,, or tension arm)
- scissors
- ruler
- two plastic planting pots, 4-inch square or equivalent size round pots
- pot saucers or dishes to hold drained water
- two coffee filters, paper towels or other liner for the pots
- potting soil: sterile, light tilth mix (approximately 1 pint)
- two to six seeds
- water (approximately 250 ml per pot)
- masking tape, sticks, pot labels, ruler

Time Requirement

This activity requires two class sessions: one class period for completion of the greenhouse and the seedbed and one class period (or less) for completion of seed starting. Because students make observations of the germinating seed over several days, the exact number of days required for completion of this activity is difficult to predict.

Teaching Suggestions

The quick-build class needs to be fast-paced. Allow students one class session working in teams of three to construct the quick-build greenhouse and plant the seeds. Customize directions to the specific materials you provide.

Arrange the common tools and supplies, such as scissors, knives, staplers, adhesives, and tape, so that all students have easy access to them. Students may need reminders or demonstrations of safe and effective use. Be alert to safety during all tool and adhesive use sessions.

Predicting the length of time needed to sprout seeds may be difficult until you have had some experience with the greenhouse lab. You will need to exercise careful and flexible planning to ensure that your students make productive use of the time between planting and seed sprouting.

Student journal writing should be an important part of this activity. Most students will need frequent and precise instructions from you to engage them in such writing. Remind them that the detailed notes they take now will prove very useful in later activities.

Before the end of class, each student should make a rough sketch of his or her team's quick-build and, for homework, should fully label each part of that sketch. If appropriate for your class, you might recommend that manual drafting, computer-aided design (CAD), or other technical illustration processes be used for later inclusion in the team report and individual portfolios.

Part IV

3 DAYS

ACTIVITY 3 Research

OVERVIEW: RESEARCH

You have completed the quick-build and observed conversion of light energy into heat. To successfully optimize this device, you will need to learn about both the physical and biological systems involved. You will also need to collect data on variables that are important to productive plant growth.

There are different ways to go about this. Trial and error is one way—fiddling around until something works. However, when you are limited in time, money, and available materials, the trial-and-error method is usually too costly to pursue.

A systematic approach to research involves careful planning and documentation and has several important strengths. Planning allows you to make reasonable cost estimates, devise a manageable time schedule, and establish a sensible division of labor. Estimates and schedules are essential if you want to finish a project on time and at a cost you can afford.

Scope of Work

- Research the biology of plant growth.
- Experiment with your quick-build greenhouse to determine baseline data.
- Research the science concepts related to the functioning of the quick-build.
- Study the physics of heat and light.
- Think critically about alternative ways to create an enclosure to provide optimum conditions for your plant.

You will probably revisit the research activities several times, because as you work through the development activities you will find that you need information you had not previously thought of. No matter how carefully you plan, you may make mistakes; be alert to learn from them and adjust your plans accordingly.

Species Specifications

Growing Pumpkins

Growing Area

It takes 120–150 days to grow an Atlantic Giant pumpkin. If you live in a region that barely experiences 120 frost-free days, you have a particular challenge to maximize the time to optimize the conditions for your plant. This means carefully planning and taking appropriate actions at specific times in the life of the pumpkin.

Research the geographic position and climate of your growing area, and answer the following questions on the geography and climate of your area:

What is the latitude?

What is the altitude?

What is the elevation?

How many frost-free days do you have in an average year?

Approximately when does the first frost-free day occur?

When do you anticipate harvesting your pumpkin for weighing-in?

ACTIVITY 3
RESEARCH

With the information you now have, what is your strategy for starting your Atlantic Giant pumpkin seeds? Describe when and how you will start your seeds, and explain why you plan to do it that way.

Growth Stages

The stages in the life cycle of a pumpkin can be defined by four distinct stages of growth, each with specific needs. These stages are the seed germination stage, the seedling growth stage, the leaf and root growth stage, and the fruit growth stage. The table below identifies the features of your plant that indicate transition into the next phase of its growth.

Research the major needs of your pumpkin plant at each growth stage.

Growth Stage	Characteristics Marking the End of This Stage
Seed germination	Emergence of seed leaves, with the seed coating still adhering
Seedling growth	Three fully developed leaves
Leaf and root growth	Fruit on the vine for two to three weeks
Fruit growth	Ends when pumpkin is harvested

Tools

The following tools should be useful in your research:

- Books and journals—suggested topics include agriculture, plants, and gardening
- Internet searches—suggested keywords include pumpkin, giant pumpkin, vegetable garden, and giant vegetable
- Personal communications with local farmers, gardening clubs, prize-winning pumpkin growers in your area, and county extension agents

3

Research Questions

1. Working with your team, make a list of questions you will need to answer in order to raise your plant properly. Divide the responsibility for finding these answers. Use the Growth Factors Chart to help identify key questions, but do not restrict your questions to the entries given there.

2. Conduct your research to find answers to as many of your questions as possible. Indicate the resource where you found each answer.

If you also found other important factors for optimal growth, specific or nonspecific to the growth stage, list them on additional pages. Remember to indicate the source for each entry; you may need to refer back to it later.

Growth Factors Chart

Growth Stage	Duration (# of days)	Ideal Conditions for Growth			
		Air Temperature	Soil Temperature	Moisture	Nutrients
Seed germination					
Seedling growth					
Leaf and root growth					
Fruit growth					

ACTIVITY 3
RESEARCH

Tolerance and Limiting Factors

The Law of Tolerance

The *law of tolerance* states that the growth of an organism is limited by the environmental factor for which that organism has the narrowest range of tolerance. An organism grows and reproduces under varying environmental conditions, such as available sunlight and moisture. When growth and reproduction are vigorous, the conditions that enable this healthy development constitute the organism's optimum range for that variable. A plant should grow easily within this optimum range (unless some other unfavorable variable interferes).

Too much or too little of an environmental variable forces an organism to work a lot harder to maintain acceptable internal conditions. The regions on the tolerance scale just above or just below the optimum range are called *stress zones.* The greater the environmental stress on an organism, the more work it has to do to stay alive, and the less energy it has available for growth and reproduction. Still higher and lower on the scale are ranges beyond which the organism cannot survive. These are the *zones of intolerance.*

Research the range of environmental variables and the physiological response of the Atlantic Giant pumpkin seedling in the different zones.

Results of Your Research

For the seedling growth stage of the pumpkin plant, make a tolerance scale (see example on page 310) to show the range of each of the following environmental variables: sunlight, temperature, and moisture.

Example of a Tolerance Scale

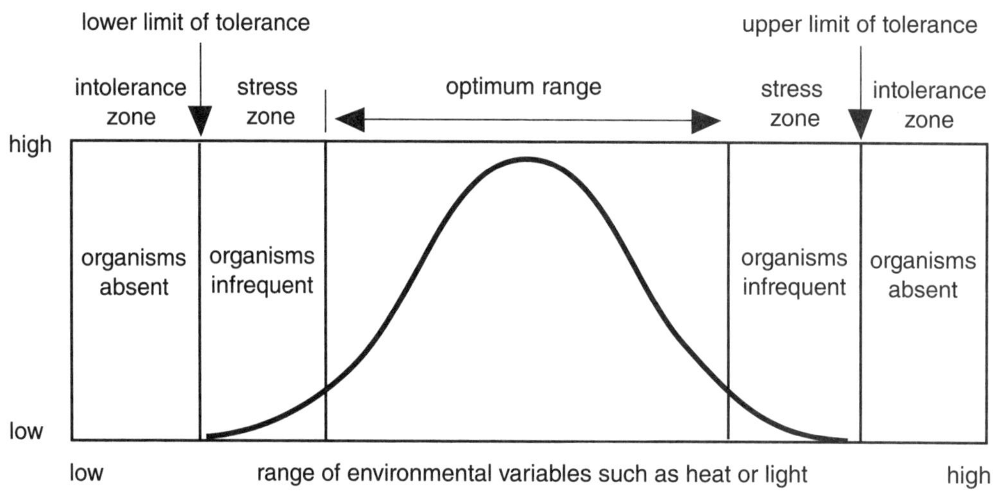

The Law of Limiting Factors

The *law of limiting factors* states that the growth of an organism will be limited if any essential growth factor is present in insufficient quantity relative to the other factors. While sunlight, temperature, and nutrients are the most important and most obvious, there are many other, less obvious, variables that may also limit growth. In an engineered environment—like your greenhouse—we can control these limiting variables to *optimize* growth conditions as the needs of the plant change from one stage to another.

Brainstorm to discover possible deficiencies and strategies to prevent, correct, or overcome them for each growth stage of the Atlantic Giant pumpkin. Then record, organize, and summarize your team's ideas.

ACTIVITY 3
RESEARCH

Identifying Variables and Making Measurements

SCiLINKS® THE WORLD'S A CLICK AWAY

Topic: scientific inquiry
Go to: *www.scilinks.org*
Code: CAH07

What Light Are We Trapping?

Depending on what is available, you can choose to use sunlight or artificial light (like a 60 W lightbulb or equivalent). Think about what is best for control and consistency, as well as what is best your for plant.

What Are We Measuring?

Before answering this question, think about what the greenhouse will be used for. Reread the Greenhouse Design Brief if you are still unclear about the main challenge. Keep in mind that you will eventually need to provide an engineered environment for at least the first two of these growth phases:

1. seed germination
2. seedling growth
3. leaf and root growth
4. fruit growth

For seed germination, collect data on these variables:

- light/heat energy supplied over time
- external air temperature over time
- internal (greenhouse) air temperature over time
- soil temperature over time

For each of these variables, think about what you have to measure and how you would go about taking these measurements. Fill in the Data Recording Plan for Seed Germination Phase.

Data Recording Plan for Seed Germination Phase

Variable	What to Measure	Frequency of Measurements	How to Measure
Energy supplied over time			
External temperature over time			
Internal temperature over time			
Soil temperature over time			

What other variables can you think of that are important for your design? Are these measurable, and if so, how would you measure them?

Repurposing for the Next Growth Stage

When your plant is at the second growth stage (seedling growth), will there be important variables to consider other than those you have just listed?

List the key differences in requirements for the first two growth stages.

ACTIVITY 3 RESEARCH

What We Do With the Data

You have already designed your experiments—now you must be sure to keep good data. But "good data" does not mean "correct answers"; "good data" means data that are carefully collected and recorded, with notes on unexpected occurrences (such as a lightbulb that burns out during the course of measurement or a greenhouse that is blown over by the wind) and deviations from the intended procedure (such as taking a reading a few minutes late or having to cut short an experiment because of time constraints).

Organize your data so that you can make sense of them and use them later when you are designing your ultimate greenhouse. Two of the most common ways to organize and represent data are with tables and graphs.

SCI LINKS THE WORLD'S A CLICK AWAY

Topic: presenting data
Go to: *www.scilinks.org*
Code: CAH08

Set Up Tables for Recording Raw and Calculated Data

Here is an example of recording and calculating data for the change of internal air temperature over time.

Time (on clock)	Elapsed Time (in minutes)	Internal Air Temperature (°C)	Total Temperature Difference (°C)
9:05 a.m.			
9:15 a.m.			
9:25 a.m.			
9:38 a.m.			
9:45 a.m.			

What questions have the sample data raised for further investigations?

Set up similar tables for all variables that you have identified and measured.

Represent Your Data Using Graphs

Here is an example for representing the change of internal air temperature over time.

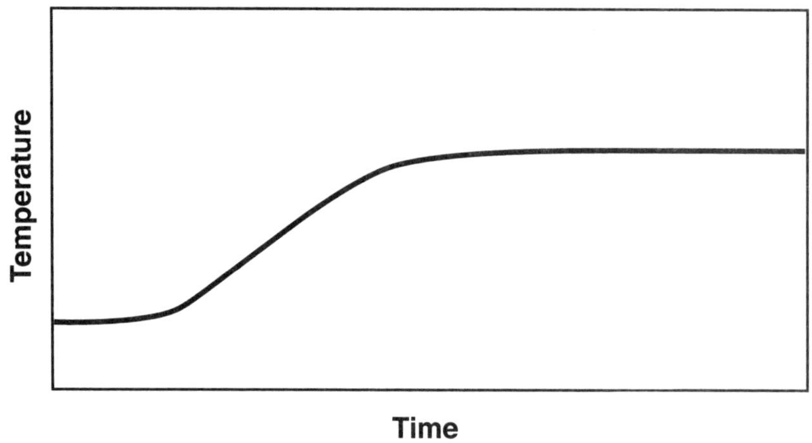

Making similar graphs for all variables that you have identified and measured will help others quickly visualize your results.

Are There Other Ways to Work With Data?

If you have access to a computer and spreadsheet software, try making the tables and graphs electronically.

ACTIVITY 3 RESEARCH

Rules of Thumb

Architects, engineers, contractors, and builders design their buildings (or other products) based on "rules of thumb" that they or other people have developed through experience. These rules of thumb tell us how something should be built to fulfill certain requirements and to serve some specific purposes.

A useful rule of thumb should be specific but not overly restrictive. You can make calculations to verify and modify rules of thumb *after* the design has been done. This means that these rules should not be taken too literally. With research and more experience, you can always revise and refine these helpful guidelines. You may have information that is more accurate or relevant to your particular situation, than these generic advisory statements.

Assignment

Whether or not a rule of thumb works well within your design depends to a certain extent on how well you understand and apply that rule. Here are several rules of thumb for designing a full-size greenhouse, taken from various sources that address a mix of horticultural as well as sunspace uses (you'll learn about sunspaces in the Passive Thermal Control activity sheet). Your assignment is to choose one rule of thumb, determine the science behind this rule, and explain how it can be applied to design your greenhouse, with or without modification. Consult the glossary for definitions of unfamiliar terms.

Some Rules of Thumb for Designing Greenhouses

Glazing

- Double glazing is recommended for latitudes above 32°N.
- The total glazing area, including overhead and vertical glass, should equal 1.5 times the area of the base of the interior.

Thermal Mass

- For storage material, it is recommended that 126–210 l of water or 600–1,000 kg of masonry be used for every square meter of south-facing glazing.
- If you can keep the soil warm, the plants will do fine, even when the temperatures are slightly below freezing.

Ventilation

- Ventilation in the summer should achieve at least one air change per minute, with inlets positioned low and outlets positioned high.
- Ventilation openings to the exterior are approximately one-sixth of floor area.

ACTIVITY 3
RESEARCH

Your Worksheet

1. What is the rule of thumb you have chosen?

2. Are there other similar rules of thumb that you have encountered? What are they? Do they support or contradict the one you have chosen from the given list?

3. Explain the scientific basis for the rule of thumb you have chosen.

4. Can this rule of thumb be applied directly to the design of your greenhouse? Explain why and how you may have to refine or modify it to serve your needs.

DAYS 3

ACTIVITY 3 Teacher Pages

OVERVIEW: RESEARCH

Students search for information on the Atlantic Giant pumpkin (or other plant you have selected) and on conditions for optimal growth. The Species Specifications activity sheet provides guidance on the questions to be explored. If you are not growing the Atlantic Giant pumpkin, use this as a model to create a specification sheet for the plant you choose. Students will use as many different types of resources as possible, including personal communication with outside experts, and share research results among teams. They will investigate the growth stages of their plant and the tolerance and limiting factors of the species, and they will brainstorm on how the law of limiting factors affects their plans.

In Identifying Variables and Making Measurements, students identify important and controllable variables for development of the greenhouse and design protocols for measuring these variables. Students consider how to represent their data in What We Do With the Data by making appropriate graphs and tables.

Students are given rules of thumb for building sunspaces and greenhouses. They examine the validity of these rules, choose one to research and study in detail, make recommendations, and suggest other rules. To review lessons on heat transfer, assign the Passive Thermal Control activity sheet in Appendix A (p. 354).

Preparation

- Provide resource lists, such as URLs, almanacs, telephone directories, and names of local farmers that you have contacted in advance.
- Provide a timeline that students can use to plan their research accordingly. Be ready to assist students with research if their seedlings emerge before they have time to complete the research activity.

ACTIVITY 3 TEACHER PAGES

Materials

For Each Student:

- Student Activity Sheets
    - » Overview: Research
    - » Species Specifications
    - » Tolerance and Limiting Factors
    - » Identifying Variables and Making Measurements
    - » What We Do With the Data
    - » Rules of Thumb
    - » Passive Thermal Control (from Appendix A)

For Each Team:

- two or three thermometers (scale 0–50°)
- hygrometer
- metric rulers
- measuring cylinder

Time Requirement

This activity requires three class sessions: one to research factors influencing the growth stages of the plant, one to brainstorm and list improvements needed in their prototype greenhouse, and one to cover the rules of thumb and have an expert on greenhouses visit the class.

Teaching Suggestions

Day 1

Group students in product research and development (R&D) teams of three to five. Review their assignment to determine factors influencing the growth stages of their plant as per the Species Specifications and the Tolerance and Limiting Factors activity sheets. Have teams work separately for 15–20 minutes, then bring the class together. Comparing findings across the class should reinforce appreciation of the numerous variables involved. List missing data and discuss where that data might be found. Have each group assign research responsibilities so that by the next class all necessary data will be in hand.

Day 2

Begin with a short team-brainstorming session based on the law of limiting factors (on the Tolerance and Limiting Factors activity sheet). After brainstorming, each team lists improvements that will be needed in their prototype greenhouse. Begin with the Identifying Variables and Making Measurements activity sheet, which includes the data recording plan, and then move one to What We Do With the Data. Be sure teams are far enough along to complete all three sheets for homework.

Day 3

Cover the rules of thumb and any incomplete business needed for the next activity, Development. This would be a good day for having a visiting expert on greenhouses. Another option is to skip this day and assign the activity sheets as homework.

Assessment

Assessment may be based on whether students state testable hypotheses, follow through and document a controlled variable experiment, graph and write interpretations of results, record reflections on areas for improvement, and contribute to their team effectively.

Part IV

DAYS 4

ACTIVITY 4
Development

OVERVIEW: DEVELOPMENT

You have completed activities investigating conversion of light energy to heat. You have also researched the light and temperature requirements for optimizing growth at various stages for your plant. To meet the Construct-a-Greenhouse design challenge, you will now develop a prototype that allows you to shift and control the balance between light absorption and light reflection, optimizing each for the changing needs of the growing plant.

Scope of Work

- Redesign your greenhouse so that it can be repurposed as plant growth requirements change.
- Build your prototype and test its performance.
- Measure and record conditions inside the enclosed environment.
- Compare growth of the enclosed plant with that of a control plant outside of the greenhouse.

Good planning is essential to good design. Take another look at the research activities for which you identified variables and wrote about the key differences in requirements for the first two growth stages. Think how to best combine design options with changing plant requirements. Be creative, but work within your objectives and constraints. Review the Inquiry Process and Design Process sheets for ideas. Evaluate your prototype critically and make modifications until you are satisfied with improved performance, or are simply out of time. Remember to reflect on your process, because how you go about your design—and what you learn—are key elements in the communication activity that follows.

SCILINKS
THE WORLD'S A CLICK AWAY
Topic: heat transfer
Go to: www.scilinks.org
Code: CAH09

Criteria and Materials

One advantage of team product development is that many crucial tasks can be split up into a manageable division of work, yielding a shortened time period for completion. From the paired trade-off factors listed on the next page (or others you can think of), divide team assignments so that each member considers only two or three, but in considerable detail.

For each factor, determine how the configuration of design criteria and materials might differ for the seed germination growth stage and the seedling growth stage. A two-column table might help you sort out the differences and trade-offs. Make notes in your laboratory journal describing your objectives and whether modifying your design is necessary to accommodate the needs of the growing plants (repurposing). Use your notes and table to formulate conclusions on the design criteria and materials to address, and describe your factor-specific recommendations in a paragraph or two. An example of these recommendations is given here for light versus heat.

Example: Light Versus Heat

The first purpose of the greenhouse is to convert incoming light to heat in order to raise the temperature of the potted soil and speed up the biological process of seed germination. From the Species Specifications information search, the soil temperature range recommended for seed germination is _________ to _________ °C.

Maintaining temperature in this range requires dark-colored, light-absorbing materials, and possibly insulation, thermal mass and/or double glazing to prevent heat loss at night. The number of days gained in the growing season will optimize chances for overall maximum growth.

After the seed leaves break ground, light absorption for conversion to heat becomes less important, but you would want to increase interior light levels to maximize photosynthesis. This repurposing will require modification of the original design, and interior surfaces exposed to incoming light should be predominantly white or reflective rather than black.

ACTIVITY 4 DEVELOPMENT

List of Trade-Off Factors

- Ventilation for temperature stability versus gas exchange
- Nutrients required at one growth stage versus nutrients required at a different growth stage
- Interior versus exterior weatherproofing
- Insulation versus accessibility and ventilation
- Strength versus cost and ease of construction
- Portability versus permanence
- Size to maturity versus material and space constraints
- Controlled versus natural light

Accumulate and discuss the input from each team member, then prioritize, vote, or otherwise reach consensus to determine your approach and choose among alternatives. Next, generate a cost estimate of materials required for your prototype. See the Scaling and Costing activity sheet for additional guidance.

4

Scaling and Costing

When you have decided on your objectives and the criteria that your design must meet, combine those factors with your research on the growth of the plant species you are working with. Your goal is to arrive at a suitable size for your prototype growth-optimizing chamber. If your goal includes growing the specimen to maturity, you must plan for the appropriate time for transplanting.

Sketching or drawing drafts of your design will help your decision-making process. You can then label parts, get an idea how the patterns could be made from available materials, and revise to economize if necessary. Compile your specific bill of materials and use telephone or e-mail inquiries, advertisements, or printed price lists to arrive at a total cost to construct. Be sure to list all your pricing sources and compare with other teams for mutual advantage.

The following partial list may help you get started; spreadsheet software (if available) could make price extensions, totaling, and revisions easier to manage.

Part	Material	Dimensions	Unit Price	Extended Price
Glazing				
Glazing frame				
Base				
Side walls				
Thermal absorber and storage				
Planting bed				
Reflecting surface				
Access opening				
Venting				
			Total cost	

ACTIVITY 4 DEVELOPMENT

Design Drawing

When you have decided your objectives and the criteria that your design will meet, you should make another sketch of the overall design and then detail the various parts, giving them names for ease of referral. On your drawing indicate with dimensions or notes the sizes of parts and how they are joined. Use the rest of this page for initial sketches and attach your prototype drawing as a separate sheet. Use of computer-assisted drawing CAD software makes initial input somewhat slower but makes revisions easier, which may pay off if you later think of design improvements.

Prototype Construction

Design Details

As you construct your prototype, make notes about the difficulties you encounter and your ideas for improvements, so that you can revise or make recommendations for others. Some sample questions about design details are given below as a checklist for your thinking:

- Is there enough height in the structure for the growing seedling?
- What is the angle of the glazing with respect to the light source?
- Is there adequate ventilation for heat and humidity control and gas exchange (following what rule[s] of thumb)?
- Will parts of the structure shade the interior at any time?
- Is the interior able to withstand high humidity?
- Do the joints prevent air leakage?
- Are there provisions for removing the grown plant from the structure?
- Is there access for watering?
- If you are making an outdoor structure, can it withstand wind and weather?
- Is there provision for protection against pests?
- Is the insulation or thermal mass adequate for temperature stabilization?
- Are light absorption and conversion to heat optimized for the germination phase?
- Is light reflectance optimized within the enclosure for the seedling phase?
- Are there any semiautomatic maintenance plans, such as a wick for watering or a passive vent opening and closing?
- Have parts been planned and laid out for cutting to produce the least material waste?
- Are the joints structurally sound for the level of use?
- Have costs been carefully calculated—are you within budget?
- How is the planting bed contained and drained?

ACTIVITY 4
DEVELOPMENT

Additional Process Questions

- Does each team member have a responsibility for specific tasks of construction? Will you employ second-party inspection and sign-offs for each aspect for *quality assurance*?
- How does your construction *schedule* coincide with the time of seed planting and sprouting?
- What kind of fertilizer *inputs* will you use, and how and when will you add them?
- What kinds of *tool(s)* will you use?

Design Improvement Notes

In the space below and/or on additional pages, organize and record your notes for design improvements. Prioritize these changes with respect to importance and feasibility. Include sections for categories such as construction rationale, difficulties, and recommendations.

Performance Tests

Preoperational Testing

To minimize potential problems and gather input for last-minute design changes, it is a good idea to pilot test your prototype before placing the sensitive biological specimen inside. Depending on how different your team's prototype is from your previously tested quick-build, you may need to measure parameters like internal temperatures and make qualitative observations on light levels, condensation, vent and access panel operation, and other factors.

Record the parameters you will measure in the space below.

Operational Monitoring

When the team is satisfied that conditions within the prototype environment will optimize growth conditions, place the specimen inside. Immediately record the starting conditions of the plant and the external as well as internal environment. These data should form the start of a periodic record of measurements and observations that will help you assess physical and biological status and make necessary adjustments.

Record the parameters you will measure in the space below.

ACTIVITY 4 DEVELOPMENT

Proof-of-Design Performance Evaluation

As indicators of successfully meeting the Construct-a-Greenhouse design challenge, there will be both biological and physical factors to consider. Your data should show that the engineered environment met specifications over the extended period of seedling growth as you anticipated. In addition, you can make both quantitative and qualitative observations of plant vigor and compare your specimen with those grown by other teams who approached the optimization challenge differently. Possible parameters of plant vigor might include leaf color, number of leaves and buds, total leaf area, and girth and height of main stem. Several solutions may produce similar results but differ in cost, appearance, ease of construction, and other factors. Perhaps some particular conclusions will emerge as universal or unique. There are many ways to achieve success.

In the space below, record the parameters that your class agreed to use to measure success.

ACTIVITY 4 Teacher Pages

4 DAYS

OVERVIEW: DEVELOPMENT

After reviewing the Design Process sheet (p. 359) and the Greenhouse Design Brief from Activity 1, students list elements of the engineered environment that are important for plant growth optimization, using the Criteria and Materials activity sheet. They compare differences and trade-offs involved in repurposing the environment from one that is aimed at accelerating germination to one that enhances the growth of the sprouted seedling. Next, students consider scaling and costing. You will need to define the limits on size and cost and provide that to students ahead of time. Alternatively, you can define the materials available and supply those materials.

Students make initial sketches based on the criteria they have identified, name parts, note dimensions, and produce a design drawing of the prototype.

In the prototype construction phase, students reflect on interlinked variables and their complexity, consider trade-offs, and make choices for design optimization to accommodate change. Depending on the scale and quality of construction feasible in your situation, this phase may take from one day to several weeks.

Students build, evaluate, and revise their greenhouse prototypes. They collect data and make quantitative and qualitative comparisons between growth of seedlings in the chamber and a control group using the Proof-of-Design Performance Evaluation activity sheet.

Preparation

- Prepare the fabrication zone.
- Consider quantities of work space, materials, and tools available, and optimize spatial arrangements for safe and efficient access.
- Set up the performance testing area. If this area will not be in your classroom, you may need to conduct careful negotiations with others in your school to ensure that the area—whether elsewhere in the school building or outdoors on the school grounds—remains undisturbed. Do not forget to include the custodian in these discussions.

ACTIVITY 4 TEACHER PAGES

- If space allows, give each group its own separate area to test the performance of its prototype.
- Consider which technological tools you might add (e.g., CAD).
- If your students are proficient with using computerized temperature probes, spreadsheet software, or graphing calculators, the data they collect during performance testing can be entered, manipulated, and printed.

Materials

For Each Student:

- Student Activity Sheets
    - » Greenhouse Design Brief (from Activity 1)
    - » Inquiry Process (from Appendix A)
    - » Design Process (from Appendix A)
    - » Overview: Development
    - » Criteria and Materials
    - » Scaling and Costing
    - » Design Drawing
    - » Prototype Construction
    - » Performance Tests
    - » Proof-of-Design Performance Evaluation

For Each Team:

- paint: white and black
- foam board, plywood, or other rigid sheet material to meet the interior or exterior conditions you specify
- thermometers (0–50°C)

- light source: direct sunlight or 100 W (or equivalent) lamp and electric outlet
- glazing: glass, Plexiglas, or clear sheet plastic
- clock or watch
- duct tape
- hammer and nail assortment
- screwdrivers and screw assortment
- waterproof glue(s)
- pencil and paper
- disposable camera, with flash, or digital camera (optional)

Time Requirement

This activity will take a minimum of four days. Define a time scale that is tight but realistic for the size and complexity of the structures you expect students to build. A simple cold frame could take one day, but a high-tech greenhouse with automatic controls might require weeks and cost a substantial amount.

Teaching Suggestions

Discussion of Homework: Design Process (optional)

Discuss the homework in the Design Process sheet. A good initial question to pose might be, *Does the diagram depicting the design process reflect your experience in designing and building?* Other questions to ask are ones from their homework:

- *Which elements of the process have you already experienced?*
- *Where in the process do you think you are now?*

You may wish to make the criteria you will use to assess students' technological design capability available at this point and invite discussion.

Suggest to students that if they become stuck at any point, they should refer again to their portfolio sheets for ideas about what to do next. Highlight, or elicit, examples of how students already used or might soon use the design process loop in nonlinear

ACTIVITY 4 TEACHER PAGES

progression. Students may wish to record scenes illustrating their group process with photographs or video.

Starting to Redesign the Improved Greenhouse

Students may choose to proceed with the redesign of their greenhouse either by improving upon their quick-builds or by starting fresh. In either case, they should (a) preserve or build on modification conclusions reached working with the quick-build system in the research phase, or (b) begin anew on a distinctly different system, in which case initial research must be inserted in the development process. Advise them also to keep the selected or specified purpose of their greenhouse constantly in mind to guide decisions.

Resources for Redesigning

Before students begin redesigning, they might individually review and then discuss as a team the Criteria and Materials activity sheet to define the team's design issues.

Timely Feedback

As students move beyond their quick-builds, it is essential that teams be allowed to grapple with the challenges of designing, building, testing, and revisiting these processes. It is important, however, that they receive input at critical times. One source of input is the team itself; encourage brainstorming as an official design process to be performed by the rules (see Design Process in Appendix A).

Another important source of feedback is timely intervention from you. Very often, a suggestion about how to solve a problem, or a reference to a diagram or picture, will stimulate action. Drawings or some sample student-built prototypes from previous classes may be helpful to stalled teams. But judgment must be exercised to avoid intervening too early.

Scale and Cost

You will need to provide limits to the scale and cost of each greenhouse. Students who find that costs are too high should be encouraged to rescale the size of their design to fit the budget. This can provide important insights on differences in the ways in which volume and surface area are scaled.

Collecting Data

Have students pretest the temperature behavior of their greenhouses before planting.

Deciding How Good Is "Good Enough"

This decision can be left to the great pumpkin. If plants survive, the system is good enough. If plants grow robustly, the design is superior.

Design Self-Assessment

As a final development activity, students reflect on what they have done thus far in the Proof-of-Design Performance Evaluation.

Assessing Students' Design and Build Capabilities

There are several key elements of students' design and build capabilities you could assess throughout the development phase:

- *How well students are able to develop solutions.* Developing ideas through to workable solutions is at the core of technological design. Look for evidence of students' ability to do this, both in your ongoing observations of students at work and in the responses they give on the activity sheets.
- *How well students are able to evaluate the processes they have used.* This includes the extent to which they are able to identify strengths and weaknesses of their prototypes. Look for evidence of students' ability to do this, both in your ongoing observations of students at work and in the responses they give on the activity sheets.
- *To what extent do students exhibit ownership of the task? Did it change with time? How much initiative did they take?* Look for evidence of this in your ongoing observations of your students.

Part IV

3 DAYS

ACTIVITY 5
Communication

OVERVIEW: COMMUNICATION

You have met the challenge. You have redesigned, built, and tested a device for growing giant pumpkins. But you will not win any prizes if nobody knows about what you have done. Until you communicate the importance of your work to the people who matter, your efforts will have little meaning. Indeed, one of the most critical abilities today is to be able to communicate clearly, effectively, and persuasively.

As an engineer, scientist, and product developer, you are dependent on funding from private and public foundations. It is important to convince those with money why they should give support to the work you do. You may want to present your ideas to local officials or create a web page that would make your work accessible to the entire world. There are agricultural fairs to show off farmers' best work, trade fairs to present new products to the market, and scientific journals to describe new discoveries. Publications are also important for communicating your findings to the greater scientific community.

Think about the discoveries you have made with the Construct-a-Greenhouse design challenge. Communicate the important parts of what you have done to several different audiences, including your classmates and a group of novice pumpkin growers. Think about the interests your audience might have in your work. Recognize that there are many possible formats to communicate with your audience. Select one or more formats to present your work.

Scope of Work

- Present your redesigned greenhouse, including the rationale, substance, and outcomes of your effort.
- Serve on the answer panel of a FAQ (frequently asked questions) site.
- Write a set of instructions for others who might be interested in building their own greenhouses.
- Write a set of "next steps" for someone to continue nurturing your plant to maturity.

5

Presentation

In Construct-a-Greenhouse, you designed and built an engineered environment for growing a giant pumpkin. You have kept careful records (perhaps even photographs), and you have the prototype model as well as the growth-enhanced biological specimen for display. You are now asked to present orally the rationale, substance, and outcomes of your effort. You have limited time for your presentation, so you should do some careful planning and rehearsal. This should be a team effort, with each team member responsible for communicating a key part of the presentation. You should also expect to field questions from your teacher and your audience. You may find visual aids useful in presenting key data and in providing your audience with tools for quick and clear analyses.

Prepare an outline of key points to cover. Focus on capturing your audience's interest, and clearly identify the strengths and distinguishing features of your design for comparison with those in other presentations. Present the evidence that the design accomplished its purpose by quantitative and qualitative comparisons with the class control plant specimen, using the criteria developed previously by the class.

List your preliminary presentation outline with team member assignments below. Consult with your teacher as a resource in your planning.

Presentation Outline	Team Member Assigned
Title:	
I.	
II.	
III.	
IV.	

Part IV

ACTIVITY 5
COMMUNICATION

Frequently Asked Questions

Your team and other teams in your class will rotate serving as an answer panel for an actual or mock FAQ site, responding with concise answers by computer, telephone, or writing. This simulates challenges faced by master gardeners consulting to the community, cooperative extension agents, pumpkin growers, or gardening columnists. To prepare well for your role, you will have to study as you would for a random question test, because you are accountable for anything covered in this project. When you don't know an answer, you should be able to point the questioner to a proper source. On the internet, there are some FAQ links on giant pumpkin pages with ready-made questions. Some of the questions and topics are listed below.

General

- How early should I start?
- How do I enter a contest?
- Are there any good books or other sources of information?
- Who holds the world record for largest pumpkin grown?
- Where can I buy good seeds?
- Which vine should I choose?
- Should I milk-feed my pumpkin?

Growing Tips

- Watering
- Mounding

Bug Problems

- Cucumber beetles
- Vine borers

Plant Problems

- Split stems/stem stress

- Yellow stem
- Deformed leaves
- Enormous seed leaves
- Plant stress
- Leaf wilt
- Variation in leaf color
- Hailstorm

Propagation

- Heating cables
- Ungerminated seeds
- Growing tips
- Seed pollination
- pH levels
- Starting seeds indoors

Fertilizer

- Fertilizer program
- Best times to fertilize the pumpkins

Weight

- Weighing process
- Weight table (over-the-top method)
- Weight of a pumpkin

Estimation

- How much does your growing pumpkin weigh?

Part IV

ACTIVITY 5
COMMUNICATION

Reflection and Recommendations

Set forth at least eight prioritized rules of thumb, factors to consider, design criteria, or statements of general wisdom and advice you would offer to others designing a greenhouse. Make your recommendations based on the experience and knowledge you have gained through research, development, and testing of your own engineered environment and on comparing your results with those of other teams.

Optimizing Germination and Seedling Growth for Transplanting

1.

2.

3.

4.

5.

6.

7.

8.

Growing a Seedling to Maturity

1.

2.

3.

4.

5.

6.

7.

8.

ACTIVITY 5 COMMUNICATION

Estimation: How Much Does Your Pumpkin Weigh?

Even if you carefully place a developing pumpkin on a bathroom scale, the increasing size and weight will quickly outpace your measurement system, and the risks of damaging the vine, scarring the rind, or injuring yourself rise accordingly. You may be inventive enough to devise an actual weighing system, but such a system would be more likely site-dedicated rather than portable, and the final weight will still be that determined at an official contest site.

Without some agreed-upon measurement system, however, growers would not have much basis other than guesswork to compare the "fruits of their labor" before the end-of-season weigh-off date. Most growers do rely on experienced judgment, but combined with either of two estimation techniques copyrighted by the World Pumpkin Confederation. Leonard Stellpflug (1989) and Howard Dill are the growers who originated the two very different estimation systems.

Stellpflug's Circumference Equation

$$\text{Weight} = 0.000213 \times \text{circumference}^3 + 20$$

The largest circumference of the fruit is measured in inches, generally parallel to the ground at approximately stem height. A table of calculated weights (in pounds) for different circumferences makes the system quite convenient. The method incorporates consideration of fruit diameter and density, including allowances for wall thickness and the seed cavity.

Starting and ending values for the actual system table are provided in the table on page 342. Calculate and provide two other values for points of your choice in between.

Circumference (inches)	Weight Estimate (pounds)
70	93
169	1,048

How do you suppose the decimal factor, cubic exponent, and addition of 20 units might have been derived?

Rewrite the equation for a system based on metric units of measure.

Dill's Over-the-Top Method

This method requires more measurements but has proved to be reliable.

1. Measure the distance over the top from ground to ground along the axis from stem to blossom end. (Allow the measuring tape to drape to the ground from the sides of the pumpkin; do not hug the undersides to the point of ground contact.)
2. Add the distance measured from ground level over the top to ground level on the opposite side at the widest portion of the pumpkin and perpendicular to the stem-blossom-end axis measured above. (Again, allow the measuring tape ends to drape.)
3. Add the circumference measured parallel to the ground from blossom end to stem.

ACTIVITY 5
COMMUNICATION

4. Multiply the sum of the three measurements by 1.9, yielding an estimate of pumpkin weight.

Study Dill's method and explain why you think it works. It may be helpful to make a drawing and consider the pumpkin first as a geometric solid. Then consider appropriate correction factors for irregularities in shape and nonuniform density. Respond in the space below.

SNAPSHOT OF UNDERSTANDING

What I Now Know About Light Absorption and Energy Conversion

1. List four key physical and natural science topics relevant to the greenhouse design challenge and demonstrate your knowledge of each in a brief paragraph. Use a separate sheet of paper if necessary.

2. Briefly describe the design process.

3. How is the design process similar to the process of scientific inquiry? How are they different?

4. Describe *quantitatively* two key parameters you controlled to achieve accelerated seed germination.

ACTIVITY 5
COMMUNICATION

5. Describe qualitatively two key factors you enhanced to optimize seedling growth.

6. What do you recommend finding out about a rule of thumb before faithfully applying it in important design work?

7. Is green a good absorber of radiant energy? Why or why not?

8. Compare and distinguish between the horticultural "greenhouse effect" experienced in this design challenge and the global environmental greenhouse effect.

3 DAYS

ACTIVITY 5 Teacher Pages

OVERVIEW: COMMUNICATION

Students, working in teams, prepare an oral presentation to communicate their R&D efforts and results to classmates and invited audience.

Students prepare written recommendations for next steps to support continued optimum growth of their plants. Beyond the end of the curriculum unit, you might encourage individual students to transfer their responsibility to other students (such as an elementary school science class) or continue as horticultural caregivers, following their own advice.

Students prepare for final assessment by assembling their portfolios according to the provided checklist, and responding to a set of frequently asked questions compiled from classmates' and others' inquiries.

Students answer Snapshot of Understanding questions to self-assess their learning.

Preparation

- Make your evaluation rubric available to students for use in presentation planning (if you have not already done so).
- Provide the highest-technology visual presentation aids feasible for student use.
- Consider offering spectator invitations and a competitive aspect.
- Review student options for extracurricular continuation of growth optimization activity. Be prepared to suggest opportunities and sources of support for various requirements, such as growing location(s), mentors, tools, water supply, soil amendments, and mulch. Some research on area agricultural fair and contest weigh-in dates might offer inspiration.

Materials

For Each Student:

- Student Activity Sheets

ACTIVITY 5
COMMUNICATION

- » Overview: Communication
- » Presentation
- » Frequently Asked Questions
- » Reflection and Recommendations
- » Estimation: How Much Does Your Pumpkin Weigh?
- » Snapshot of Understanding

For Each Team:

- visual presentation system (marker board, overhead projector, etc.)
- oral presentation assessment rubric

Time Requirement

This activity requires three class sessions:

- one class for team formulation of reflections and recommendations and preparation of presentation, with additional time needed for homework;
- one and a half classes for team presentations and FAQ preparation; and
- half a class for final evaluation and self-assessment.

Completion of the final Snapshot of Understanding takes about 35 minutes.

Teaching Suggestions

Reflection and Recommendations

Students may wish to continue nurturing their specimen beyond the scope of the unit and the capacity of their prototype, by transplanting it outdoors to a home or school garden, competing in a fair or contest, or growing it for market, decoration, or further learning. Any such initiative and follow-through clearly merits encouragement and support for furthering student academic or hobby/sport enjoyment. In any case, written continuation recommendations and instructions are the assigned product from the reflection activity, and this product should be evaluated for its basis in research and realistic projection on the part of all team members.

Presentations

Teams are asked to divide presentation responsibilities among members so that effort and benefit of the work and learning are distributed equitably. The team should provide emotional and academic support, where needed, with particular sensitivity to ensuring to the greatest extent possible that this assignment in public speaking be a positive experience for each member. An assessment rubric, tailored to fit your class learning and evaluation objectives, should be made available to teams well ahead of time (possibly at the beginning of the prototype development phase) to encourage maximum awareness, planning, and preparation.

ACTIVITY 5
TEACHER PAGES

Sample Oral Presentation Assessment Rubric

Evaluation Criteria	Points
Presenters/Participations	5
Date/Time/Length	5
Title/Topic/Communication of Objective	10
Methods Used/Style/Visual Aids	20
Interaction/Engagement With Audience	20
Delivery Factors	10
Goal Achievement	10
Strengths/Effectiveness	10
Areas to Improve—Self-Assessment	10
Total	100

Overall Checkpoints

Is the information presented

- relevant?
- purposeful?
- organized?
- accurate?
- interactive?

Presenters' Report

Did the presenters report

- science principles that explain observations?
- what they would do differently again and next?
- how they employed processes of inquiry and design to meet the challenge?

Completing the Snapshot of Understanding

Warning: This Snapshot has an important catch! By using thin plastic wrap they lose the greenhouse effect. Contrary to expectations, students should not observe much temperature gain. You will learn a great deal about their understanding of the greenhouse effect by observing how they attempt to answer the questions in this Snapshot and relate them to the earlier Snapshot.

After students complete the final Snapshot of Understanding, provide a brief time for them to compare their new answers with those on their prior-learning Snapshot. Seeing how much is learned in this brief, but intensive, experience can be informative.

APPENDIX A
Side Roads

The material in this appendix is intended to support activities that you may choose to add to those described in the core unit. Many of these are key activities, but they have been placed in this appendix because they can fit in several different places in Construct-a-Greenhouse—exactly where they are used is a matter best decided by you in response to student questions and feedback. Some activities may be profitably used more than once. An analysis of the design process, for example, will provide different insights when used in the research activities than when used in the design activities.

In this appendix:

- Inquiry Process
- Passive Thermal Control
- Design Process

A

Inquiry Process

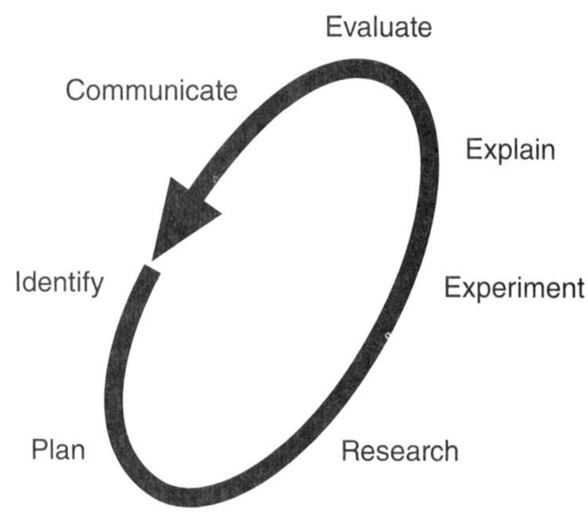

The inquiry process is often viewed as a cycle of action that repeats until the investigators reach a satisfying solution. It can be described with seven basic elements:

- *Identify* and clarify questions. Understand the issue or problem, and make a testable hypothesis.
- *Plan* appropriate procedures. Brainstorm, draw and write ideas, clarify their ideas, and suggest possible strategies or methods.
- *Research* major concepts. Learn what is known about the situation from sources other than actual investigation, and obtain information from preliminary experiments. Decide what technology, approach, equipment, and safety precautions are useful. Document your experiments and log your data.
- *Experiment*. Use tools and measuring devices to conduct experiments. Use calculators and computers to store and present data.
- *Explain* logical connections. Analyze your data. Formulate explanations using logic and evidence, and possibly by constructing a physical, conceptual, or mathematical model.
- *Evaluate* alternatives. Compare your explanations to current scientific understanding and other plausible models. Identify what needs to be revised, and find the preferred solution.
- *Communicate* new knowledge and methods. Communicate results of your inquiry to your peers and others in the community. Construct a reasoned argument through writing, drawings, and oral presentations. Respond appropriately to critical comments.

APPENDIX A
SIDE ROADS

Questions

Read the following questions, but do not answer them until after your team has experienced working together on the design challenge research activities.

1. Make your own checklist of team activities that correspond to steps in the cycle described above.

2. Create your own version of the inquiry process using words and pathways that fit your team's activity.

3. What shape is your inquiry pathway diagram (circle, spiral, cascade, other)?

4. How and where do the seven steps described above fit within your process description?

A

SCiLINKS. THE WORLD'S A CLICK AWAY

Topic: heat transfer
Go to: *www.scilinks.org*
Code: CAH09

Topic: conduction/convection
Go to: *www.scilinks.org*
Code: CAH01

Topic: radiation
Go to: *www.scilinks.org*
Code: CAH02

Passive Thermal Control

Heat Transfer Processes

When solids, liquids, or gases of different temperature come into contact, without further disturbance, heat flows from warmer to cooler entities to reach a common temperature. Three basic heat transfer processes are summarized in Figures 1–5: conduction, convection, and radiation.

A material may achieve equilibrium with its surroundings through any or all of the processes.

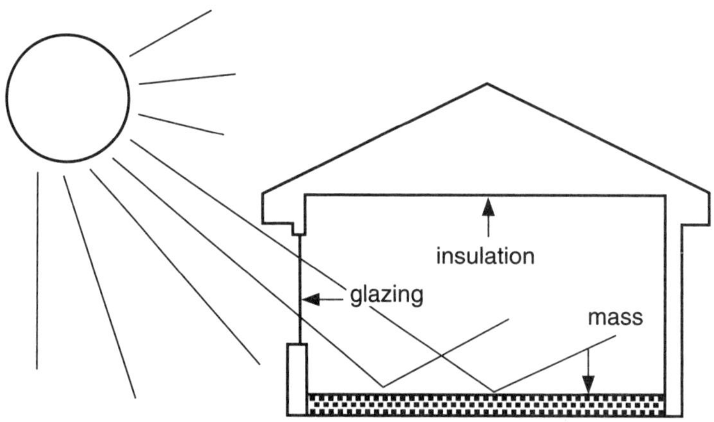

Figure 1. Sunlight enters a building directly through windows. Large amounts of solar heat can be absorbed and stored by heavy materials in the building for later use.

APPENDIX A
SIDE ROADS

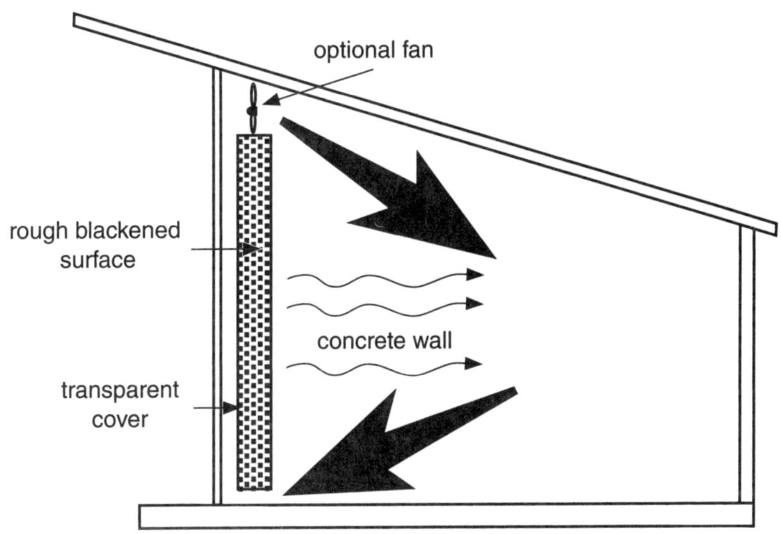

Figure 2. The thermal storage wall is made of heavy materials such as concrete, stone, or containers of water. Radiant energy that passes through the windows is stored in the wall. At night, the wall radiates warmth to the building. With vents at the top and bottom of the wall, some of the heat entering the building is circulated directly by natural convection.

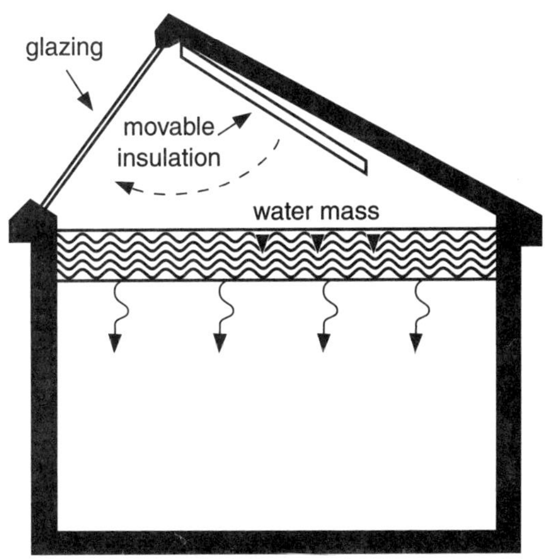

Figure 3. Most thermal storage roofs consist of waterbed-like containers. The water absorbs the radiant energy from the Sun, conducts the heat energy throughout the ceiling, and warms the building by radiation.

A

Figure 4. A sunspace, such as a greenhouse or an atrium, can provide considerable solar heat to a building and reduce heat loss. Can you identify the features of a sunspace?

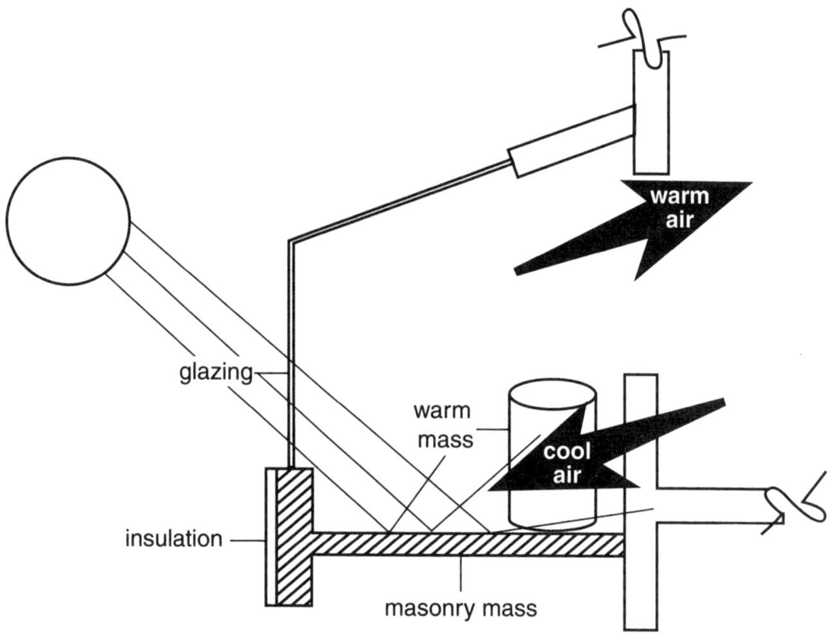

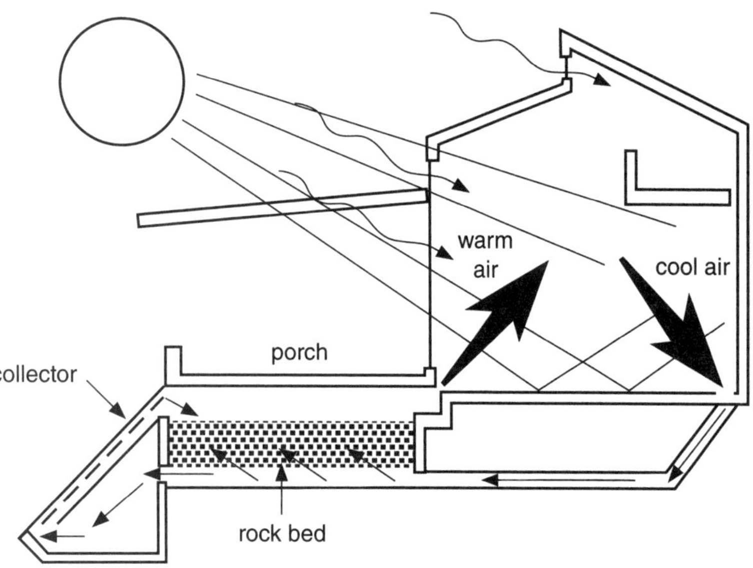

Figure 5. A connective loop is formed when hot air rises and cool air moves in to replace it. Can you describe what is happening in the building illustrated in this figure?

APPENDIX A
SIDE ROADS

Thermal Mass Materials

Each material responds uniquely to thermal (and light) energy input. Several relevant properties of materials have been described to characterize the differences:

- density
- specific heat
- heat capacity
- surface finish
- emissivity
- conductivity
- reflectance
- R-factor

In addition there are properties related to the geometry of where materials are placed. These include convection and angle of incidence.

Assignment

Divide up the parameters among your team members. Each team member should prepare a review of key concepts related to the parameters assigned to them. In addition, select those parameters that require further research and design and conduct a set of simple experiments to help your team make soundly based quantitative and qualitative choices among alternative materials, passive control strategies, and configurations for your prototype design. For example, Cheerios can make good thermal insulators, but the R-factor is not printed on the box. If you plan to use Cheerios, you will need to determine the R-factor and compare it with that of other possible materials. Summarize the following for each experiment:

- objective
- approach
- materials
- observations and measurements

- data analysis
- conclusion(s)

The methods and results of your experiments will be reviewed in the proof-of-design performance test comparisons with other teams as well as referenced in your team's final presentation and recommendations.

Part IV

APPENDIX A
SIDE ROADS

Design Process

The design process is often viewed as a cycle of action that repeats until the designers reach a satisfying solution. It can be described with seven basic elements: identify, create, investigate, choose, implement, evaluate, and communicate.

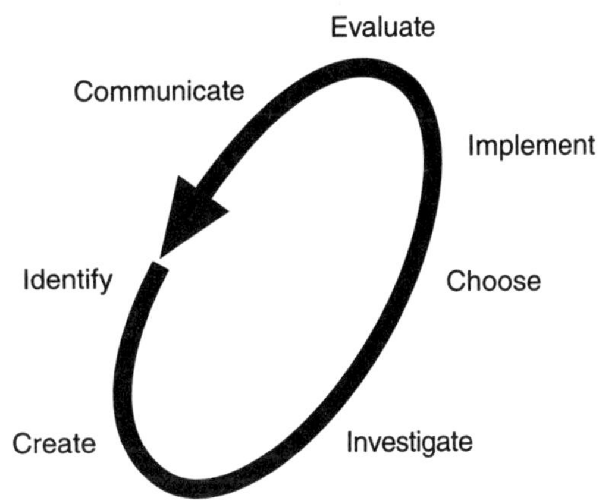

- *Identify* and clarify the situation. Understand the challenge or problem, including the criteria for success and constraints on the design.
- *Create* solutions. Brainstorm, draw and write ideas, and suggest possible strategies or methods.
- *Investigate* possibilities. Learn what is known about the situation, and what technology or approach could be useful. Conduct experiments to test your ideas.
- *Choose* a solution. List the solutions most likely to be successful, and make decisions for how well each solution meets the design challenge or solves the problem.
- *Implement* the design. Learn that a successful design often depends on good fabrication, whether it is a scaled or life-size version of the product.
- *Evaluate* the design. Perform tests to obtain the feedback that informs them about the parts of the design that worked or needed improvement.
- *Communicate* the solution. Present your designs to your peers and others in the community, communicating your ideas through drawings, writing, formal presentations, informal discussions.

A

Questions

After reading about the design process, answer the following questions:

1. Which elements of the process have you already experienced?

2. Which elements have you not yet experienced?

3. Where in the process do you think you are now?

4. What will your next steps be?

Part IV

APPENDIX B
Text Reconstruction Exercises

Energy Transfer

The three paragraphs below describe important information about energy transfer. To keep this information confidential, some of the sentences within each paragraph have been reordered. Your task is to restore them to their proper order.

¶1

___ Trust us, it will be.
___ The famous physicist Richard Feynman said that nobody really knows what energy is—all we know is that it is conserved.
___ If we keep track of all the places it goes, we note that there is always the same amount after as there was before.
_2_ Energy keeps getting transformed from one type to another.
___ Unfortunately, in most everyday situations it is very difficult to keep track of all the transformations energy goes through and so it often seems some energy is getting lost.
___ This can make learning about energy difficult and confusing unless you are willing to accept on faith that it is always conserved.

¶2

___ Thus, we have three kinds of energy to consider: light, heat, and chemical.
___ If there are green plants in the greenhouse, then some of the light is converted by photosynthesis into the chemical energy that binds organic matter together.
___ Light is one form of energy and heat is another.
_2_ In the greenhouse, light is converted into heat.

¶3

___ Because we will not be measuring, we will not need to make direct use of the principle of energy conservation.
___ In the greenhouse design, you will be considering energy transfers among these three kinds of energy.
_2_ But we will not need to measure the amount of energy transferred.
___ Still, it is useful to know that the only energy we have to work with is the light that comes through the window.

B

Repurposing

The paragraphs below describe the concept of repurposing. In this case, the paragraphs have been repurposed into a puzzle, and some of the sentences in each paragraph have been reordered. Your task is to restore them to the original order.

¶1

_4_ Think about how you might redesign a shirt so it could be both a shirt and a towel.
___ Have you ever dried your hands on your shirt or dress?
___ This is an example of repurposing.
___ Shirts and dresses are not designed to be towels, but since some shirts and dresses are designed to absorb sweat, they can make pretty good towels.

¶2

___ This conflict between our two goals is known as a trade-off.
___ The problem with a device that has more than one purpose is that it may not be possible to accomplish both purposes well.
___ Our shirt-towel probably has a long tail or other extension that makes an okay towel, but adds unnecessary bulk to the shirt.
_4_ To make a better shirt, we may have to make a worse towel.

¶3

___ Trade-offs are important in every design because we always have more than one consideration.
___ Cost and style are two considerations that affect most designs, yet are almost never completely in line with the device's main objective.

¶4

___ The process of making changes to these jelly jars to make them good drinking glasses is known as repurposing.
___ Some objects are designed to serve one purpose for a while and then a second purpose later.
_2_ For example, some jelly jars are designed to be used as drinking glasses after the jelly has been used up.
___ In Construct-a-Greenhouse, you will repurpose a device designed to convert light into heat to make it a device that concentrates light on the leaves of plants.

Part IV

APPENDIX B
TEXT RECONSTRUCTION EXERCISES

Plant Growth

The paragraphs below describe stages of plant growth. Sentences within each paragraph have been reordered. Your task is to find the original order.

¶1

___ They begin life as a small seed that is self-contained and can survive on its own for a very long time (in some cases, thousands of years).
___ Plants are amazing.
___ All the time the seed is checking to see if conditions are ready for it to sprout.
___ Sprouting is determined by temperature and humidity.
_5_ As the seed sprouts, it uses its own stored energy source (food) to grow.
___ In order to burn this energy, the plant must take oxygen from the air.

¶2

___ The chemicals are carbon, oxygen, hydrogen, and nitrogen.
___ Once the seedling has developed leaves, it begins to generate its own food.
_2_ It makes this food by converting light energy into chemical bond energy through a process known as photosynthesis.
___ This process requires light and chemicals.
___ The plant finds all these chemicals except nitrogen in the air (not in the soil, as is often assumed).
___ During the day, the plant absorbs carbon dioxide from the air. It generates oxygen as a waste product.
___ At night, or when there is little light, the plant absorbs some oxygen from the air as it burns its food.

¶3

___ As long as they have no leaves, they will not remove carbon dioxide from the air, and they will not generate excess oxygen.
___ Some plants take a rest in the winter.
___ They lose their leaves or retreat down into their roots.
_3_ These plants continue to burn a little oxygen as they "hibernate."

B

Text Reconstruction Key

The paragraphs below show the correct order of sentences for each text reconstruction exercise. When you hand out the initial homework assignment, ask students to number the sentences in each paragraph so as to put them in the correct order. It is also highly beneficial to ask students to rewrite the paragraphs once they have determined the correct order.

Energy Transfer

Paragraph 1

1. The famous physicist Richard Feynman said that nobody really knows what energy is—all we know is that it is conserved.
2. Energy keeps getting transferred from one type to another.
3. If we keep track of all the places it goes, we note that there is always the same amount after as there was before.
4. Unfortunately, in most everyday situations, it is very difficult to keep track of all the transformations energy goes through and so it often seems some energy is getting lost.
5. This can make learning about energy difficult and confusing unless you are willing to accept on faith that it is always conserved.
6. Trust us, it will be.

Paragraph 2

1. Light is one form of energy and heat is another.
2. In the greenhouse, light is converted into heat.
3. If there are green plants in the greenhouse, then some of the light is converted by photosynthesis into the chemical energy that binds organic matter together.
4. Thus we have three kinds of energy to consider: light, heat, and chemical.

APPENDIX B
TEXT RECONSTRUCTION EXERCISES

Paragraph 3

1. In the greenhouse design, you will be considering energy transfers among these three kinds of energy.
2. But we will not need to measure the amount of energy transferred.
3. Because we will not be measuring, we will not need to make direct use of the principle of energy conservation.
4. Still, it is useful to know that the only energy we have to work with is the light that comes through the window.

Repurposing

Paragraph 1

1. Have you ever dried your hands on your shirt or dress?
2. This is an example of repurposing.
3. Shirts and dresses are not designed to be towels, but since some shirts and dresses are designed to absorb sweat, they can make pretty good towels.
4. Think about how you might redesign a shirt so it could be both a shirt and a towel.

Paragraph 2

1. The problem with a device that has more than one purpose is that it may not be possible to accomplish both purposes well.
2. Our shirt-towel probably has a long tail or other extension that makes an okay towel, but adds unnecessary bulk to the shirt.
3. This conflict between our two goals is known as a trade-off.
4. To make a better shirt, we may have to make a worse towel.

Paragraph 3

1. Trade-offs are important in every design because we always have more than one consideration.
2. Cost and style are two considerations that affect most designs, yet are almost never completely in line with the device's main objective.

Paragraph 4

1. Some objects are designed to serve one purpose for a while and then a second purpose later.
2. For example, some jelly jars are designed to be used as drinking glasses after the jelly has been used up.
3. The process of making changes to these jelly jars to make them good drinking glasses is known as repurposing.
4. In Construct-a-Greenhouse, you will repurpose a device designed to convert light into heat to make it a device that concentrates light on the leaves of plants.

Plant Growth

Paragraph 1

1. Plants are amazing.
2. They begin life as a small seed that is self-contained and can survive on its own for a very long time (in some cases, thousands of years).
3. All the time the seed is checking to see if conditions are ready for it to sprout.
4. Sprouting is determined by temperature and humidity.
5. As the seed sprouts, it uses its own stored energy source (food) to grow.
6. In order to burn this energy, the plant must take oxygen from the air.

APPENDIX B
TEXT RECONSTRUCTION EXERCISES

Paragraph 2

1. Once the seedling has developed leaves, it begins to generate its own food.
2. It makes this food by converting light energy into chemical bond energy through a process known as photosynthesis.
3. This process requires light and chemicals.
4. The chemicals are carbon, oxygen, hydrogen, and nitrogen.
5. The plant finds all these chemicals except nitrogen in the air (not in the soil, as is often assumed).
6. During the day, the plant absorbs carbon dioxide from the air. It generates oxygen as a waste product.
7. At night, or when there is little light, the plant absorbs some oxygen from the air as it burns its food.

Paragraph 3

1. Some plants take a rest in the winter.
2. They lose their leaves or retreat down into their roots.
3. These plants continue to burn a little oxygen as they "hibernate."
4. As long as they have no leaves, they will not remove carbon dioxide from the air, and they will not generate excess oxygen.

APPENDIX C
Sample Answers

Species Specifications

Activity 3, p. 306

Growing Pumpkins

Students are asked to research the geographic position and climate of their growing area. Here are some suggestions for where to find the answers to the questions in this activity sheet.

What is the latitude? *local map, weather station*

What is the altitude? *local map, airport, weather station*

What is the elevation? *measure yourself*

How many frost-free days do you have in an average year? *local farmers, weather station*

Approximately when does the first frost-free day occur? *local farmers, weather station*

When do you anticipate harvesting your pumpkin for weighing-in? *you decide*

Activity 3, p. 310

The Law of Limiting Factors

Students brainstorm possible deficiencies and strategies to prevent, correct, or overcome these deficiencies, for each growth stage of the Atlantic Giant pumpkin. As students record, organize, and summarize their team's ideas, make sure they include the following factors:
Water, maximum sunlight, adequate fertilizer, and heat

Other protective factors might include:
Protect the base of the pumpkin from rot
Protect the plant from pests—insects, birds, rodents
Protect the plant from theft

Part IV

APPENDIX C
SAMPLE ANSWERS

Data Recording Plan for Seed Germination Phase

Activity 3, p. 312

Sample answers to the data recording plan are given below:

Variable	What to Measure	Frequency of Measurements	How to Measure
Energy supplied over time	*Sunlight; temperature*	*Several times a day*	*Light meter; thermometer*
External temperature over time	*Temperature in shade*	*Several times a day*	*Thermometer*
Internal temperature over time	*Temperature in standards place*	*Once or twice a day*	*Thermometer*
Soil temperature over time	*Temperature in soil*	*Once or twice a day*	*Thermometer*

Rules of Thumb: Student Worksheet

Activity 3, p. 315

1. What is the rule of thumb you have chosen?
   Total glazing = 1.5 times the area of the floor

2. Are there other similar rules of thumb that you have encountered? What are they? Do they support or contradict the one you have chosen from the given list?
   Plant after Memorial Day.

3. Explain the scientific basis for the rule of thumb you have chosen.
   We need more glass than floor, because as the Sun moves, the area lit will move. With more glass, we keep the floor fully lit for a longer time.

4. Can this rule of thumb be applied directly to the design of your greenhouse? Explain why and how you may have to refine or modify it to serve your needs.
   Yes, it helps in designing the right amount of window. The insta-build and the quick-build both have too little glass.

Glossary

absorption: when light shines on a surface, some of the light is absorbed and converted into heat energy

angle of incidence: the angle between the surface of an object and the rays (usually light) falling on the surface

baseline data: data taken to determine conditions before an experiment is started

brainstorming: a group problem-solving technique which involves the spontaneous contribution of ideas from all members of the group

condensation: when a gas converts to a liquid; water vapor will condense into liquid water drops when it comes in contact with a cold surface

conduction: the flow of thermal energy through a substance from a higher- to a lower-temperature region

convection: the movement of heat through a fluid (such as liquid or gas); for example, the free or forced movement of air throughout a room

emissivity: a measure of the amount of light or thermal radiation a warm body will emit

energy conversion: energy comes in different forms and is often converted from one form to another; heat is converted into light, light into heat, motion into heat, and so forth

energy transfer: the movement of energy from one place to another or from one form to another

equilibrium: a balance between two opposing processes such that the net effect of the two processes is no total change; for example, we breathe in about as much air as we breathe out, so over time our lungs stay roughly the same size

factors: the different elements out of which a whole object is made; the term is often used to describe the different causes that lead to a particular outcome

germination: the phase a seed goes through during which the seed changes

GLOSSARY

from a closed pod dependent on its own food to a plant with leaves, able to feed itself

glazing: the glass in a window

incandescent radiation: radiation given off from a hot surface

inorganic: material that was not originally part of a living organism (e.g., rocks or sand)

interlinked variables: a set of variables that depend on each other

nutrients: the food inputs a living organism needs to absorb to survive and grow

optimal conditions: the best situation for accomplishing some purpose

organic: material that was part of a living organism (e.g., wood)

organism: a living entity or unit; a single cell or a complex collection of cells, such as an animal or a tree

performance test: a test to determine how well a design accomplishes an intended purpose

prototype: an original model; first full-scale and (usually) functional form of a new type of construction

R-factor: a measurement system for determining the thermal insulation value of a material or object

radiation: the dissemination of energy from a source; a type of heat transfer that takes place through exposure to a series of electromagnetic waves, such as infrared waves

reflectance: a measure of how much light a surface will reflect

repurpose: to reorganize a design such that it can accomplish a task different from the one for which it was originally made

rubric: a set of rough rules or principles used in evaluating quality

GLOSSARY

scaling: a proportional increase or decrease in the size of something

specifications: the precise details of an invention, plan or proposal

temperature: a scale for measuring thermal energy by showing how warm or cold an object is relative to something else

thermal mass: the capacity of a body to absorb heat

trade-off: when one quality of a design is sacrificed to improve another; for example, to make an object lighter, it may be necessary to sacrifice strength

transplanting: the relocation of a plant from one place to another

variable: an object or quality of changeable value

ventilation: the exchanging of air in a closed space with "fresh" air from outside

Part IV

Suggested Readings

American Association of Physics Teachers (AAPT). 1986. *Solar energy, book II*. AAPT.

Driver, R. 1987. *Approaches to teaching plant nutrition.* Leeds, England: University of Leeds, Centre for Studies in Science and Mathematics Education, Children's Learning in Science Project.

Mazria, E. 1979. *The passive solar energy book*. Emmaus, PA: Rodale Press.

Schauble, L., L. E. Klopfer, and K. Raghavan. 1991. Students' transition from an engineering model to a science model of experimentation. *Journal of Research in Science Teaching* 28 (9): 859–882.

Schneider, S., and R. Londer. 1984. *The coevolution of climate and life*. San Francisco: Sierra Club Books.

Stellpflug, L. 1989 (September). *WPC Newsletter.*

Wiser, M. 1988. The differentiation of heat and temperature: History of science and novice-expert shift. In *Ontogeny, phylogeny, and historical development,* ed. S. Strauss, 28–48. Norwood, NJ: Ablex.

Index

INDEX

INDEX

INDEX